Logistical Asia

Brett Neilson • Ned Rossiter
Ranabir Samaddar
Editors

Logistical Asia

The Labour of Making a World Region

Editors
Brett Neilson
Institute for Culture and Society
Western Sydney University
Parramatta, NSW, Australia

Ned Rossiter
Institute for Culture and Society
Western Sydney University
Parramatta, NSW, Australia

Ranabir Samaddar
Calcutta Research Group
Kolkata, West Bengal, India

ISBN 978-981-10-8332-7   ISBN 978-981-10-8333-4   (eBook)
https://doi.org/10.1007/978-981-10-8333-4

Library of Congress Control Number: 2018935185

© The Editor(s) (if applicable) and The Author(s) 2018
This work is subject to copyright. All rights are solely and exclusively licensed by the Publisher, whether the whole or part of the material is concerned, specifically the rights of translation, reprinting, reuse of illustrations, recitation, broadcasting, reproduction on microfilms or in any other physical way, and transmission or information storage and retrieval, electronic adaptation, computer software, or by similar or dissimilar methodology now known or hereafter developed.
The use of general descriptive names, registered names, trademarks, service marks, etc. in this publication does not imply, even in the absence of a specific statement, that such names are exempt from the relevant protective laws and regulations and therefore free for general use.
The publisher, the authors and the editors are safe to assume that the advice and information in this book are believed to be true and accurate at the date of publication. Neither the publisher nor the authors or the editors give a warranty, express or implied, with respect to the material contained herein or for any errors or omissions that may have been made. The publisher remains neutral with regard to jurisdictional claims in published maps and institutional affiliations.

Cover illustration: © Amarin Jitnathum / EyeEm / Getty Images

Printed on acid-free paper

This Palgrave Macmillan imprint is published by the registered company Springer Nature Singapore Pte Ltd. part of Springer Nature.
The registered company address is: 152 Beach Road, #21-01/04 Gateway East, Singapore 189721, Singapore

ACKNOWLEDGEMENTS

Although the chapters in this book focus on logistics in the making of the Asian region, and in particularly the city of Kolkata, they are the fruit of a longer research project dubbed 'Logistical Worlds: Infrastructure, Software, Labour,' which was conducted across the cities of Athens, Kolkata, and Valparaíso. The editors would like to extend thanks to the authors as well as all the other researchers who participated in 'research platforms' staged in these cities. In Athens, these researchers include Ursula Huws, Anna Lascari, Ilias Marmaras, Sandro Mezzadra, Dimitrios Parsanoglou, Carolin Philipp, and William Walters. In Kolkata, they include Orit Halpern, Mithilesh Kumar, Sugata Marjit, Cecile Martin, Immanuel Ness, and Elisabeth Simbürger. In Valparaíso, they include Jamie Allen, Jorge Budrovich, Cristian Ceruti, Lucas Cifuentes, Hernán Cuevas, Katie Detwiler, Alejandro Donaire, Orit Halpern, Valentina Leal, Jose Llano, Elisabeth Simbürger, and Paulina Varas. Special thanks go to the following individuals for their role in organizing the research platforms: Pavlos Hatzopoulos and Nelli Kambouri (Athens), Mithilesh Kumar (Kolkata), and Jorge Budrovich and Cristian Ceruti (Valparaíso). The research for this book has also benefited from our involvement in the summer university on Investigating Logistics held at Humboldt University, Berlin, in September 2016. Thanks to Manuela Bojadžijev and Sandro Mezzadra for orchestrating that event.

The Logistical Worlds project was funded by the Australian Research Council Discovery Project Scheme—DP130103720 'Logistics as Global Governance: Infrastructure, Software, and Labour along the New Silk Road.' We are also grateful for funding from the Social Science Research

Council (New York) that allowed us to organize a workshop entitled 'Logistics of Asia-Led Globalization: Infrastructure, Software, Labor' at the InterAsian Connections V conference at Seoul National University in April 2016. Many of the contributors to this volume gathered at that event. We also extend thanks to participants in the Seoul workshop who did not end up contributing to this volume: Thilini Kahanawaarachi, Ping Sun, and John Walsh. The authors of Chap. 12 thank the Macau Cultural Affairs Bureau for funding their research.

Finally, we are grateful to Annee Lawrence for her exacting copyediting and Jacob Dreyer, our editor at Palgrave, for seeking out work on this project and assisting us at all steps of the publication.

Contents

1 Making Logistical Worlds 1
  Brett Neilson, Ned Rossiter, and Ranabir Samaddar

Part I Port as Infrastructure of Postcolonial Capitalism 21

2 The Port of Calcutta in the Imperial Network of South
  and South-East Asia, 1870s–1950s 23
  Kaustubh Mani Sengupta

3 Spatialization of Calculability, Financialization of Space:
  A Study of the Kolkata Port 47
  Iman Mitra

4 Ports and Crime 69
  Paula Banerjee

5 Haldia: Logistics and Its Other(s) 91
  Samata Biswas

6 Kolkata Port: Challenges of Geopolitics and Globalization 113
  Subir Bhaumik

Part II Logistics of Asia-Led Globalization 133

7 The Importance of Being Siliguri: Border Effect
 and the 'Untimely' City in North Bengal 135
 Atig Ghosh

8 Piraeus Port as a Machinic Assemblage: Labour, Precarity,
 and Struggles 155
 Pavlos Hatzopoulos and Nelli Kambouri

9 Asia's Era of Infrastructure and the Politics of Corridors:
 Decoding the Language of Logistical Governance 175
 Giorgio Grappi

10 Logistics of the Accident: E-Waste Management
 in Hong Kong 199
 Rolien Hoyng

11 Geopolitics of the Belt and Road: Space, State,
 and Capital in China and Pakistan 221
 Majed Akhter

12 Becoming Immaterial Labour: The Case of Macau's
 Internet Users 243
 Zhongxuan Lin and Shih-Diing Liu

13 Follow the Software: Reflections on the Logistical
 Worlds Project 263
 Brett Neilson

Index 291

NOTES ON CONTRIBUTORS

Majed Akhter is a lecturer in the Department of Geography at King's College London. His primary research interest is in state formation, large physical infrastructures, and the uneven geographies of capitalist development. His research appears in journals such as *Antipode, Critical Asian Studies, Political Geography, Geoforum, Progress in Human Geography*, and the *Transactions of the Institute of British Geographers*.

Paula Banerjee is a professor in the Department of South and South-East Asian Studies at the University of Calcutta and Director of the Calcutta Research Group. She is best known for her work on women in borderlands and women and forced migration. Her recent publications include *The State of Being Statelessness* (edited, 2016), *Unstable Populations, Anxious States* (edited, 2013), *Women in Indian Borderlands* (edited, 2012), and *Borders, Histories, Existences: Gender and Beyond* (2010).

Subir Bhaumik is a veteran BBC correspondent now working as consulting editor with Myanmar's Mizzima Media Group. He is a former Queen Elizabeth House Fellow of Oxford University and a Eurasian Nets fellow at Frankfurt University. He has written two books on regional issues and geopolitics and edited three volumes, besides contributing articles to volumes edited by leading scholars.

Samata Biswas teaches English literature at Bethune College, Kolkata, India. Her doctoral research was about body cultures in contemporary India, analysing fitness, weight loss, and diet discourses as present in popular media as well as through narratives of participants. She is interested in visual culture, gender studies, and literature and migration. At present, she is trying to map Kolkata as a sanitary city, focusing on access to clean sanitation or the lack thereof. She runs the blog *Refugee Watch Online*.

Atig Ghosh is an assistant professor at Visva Bharati University and a member of the Calcutta Research Group. He is the author of *Branding the Migrant* (2013) and *The State of Statelessness* (edited, 2017).

Giorgio Grappi is a research fellow at the University of Bologna, Department of Political and Social Sciences. He is interested in the transformation of the state form and the reconfiguration of political scales. He has been involved in the project *Logistical Worlds: Infrastructure, Software, Labour* and is now part of the EU-funded Horizon 2020 project GLOBUS (http://www.globus.uio.no/). His most recent publications include 'Beyond Zoning: India's Corridors of Development and New Frontiers of Capital' with Ishita Dey (*South Atlantic Quarterly*, 2015) and the book *Logistica* (Ediesse, Roma, 2016).

Pavlos Hatzopoulos is affiliated to Panteion University, working on the digitalization of labour and migration. He was part of the international project *Logistical Worlds: Infrastructure, Software, Labour*, conducting research on Chinese investments around Piraeus port and focusing on the optimization of production, labour relations, and technologies of governance. Hatzopoulos is based in Athens.

Rolien Hoyng is an assistant professor in the School of Journalism and Communication at the Chinese University of Hong Kong. She holds a PhD from the University of North Carolina at Chapel Hill, USA, and MA degrees from the University of Amsterdam, the Netherlands. Her work is primarily situated in Istanbul and Hong Kong. It explores digital infrastructure in relation to urbanism as well as digital culture. Her recent articles have appeared in *International Journal of Communication*, *European Journal of Cultural Studies*, *Television and New Media*, and *Javnost-The Public*.

Nelli Kambouri is a gender scholar who has been working at the Center for Gender Studies in the Department of Social Policy of Panteion University in Athens since 2005. She teaches gender, labour, and social policy and conducts research on gender and logistics. Kambouri lives and works in Athens.

Zhongxuan Lin is an associate research fellow at the School of Communication and Design, Sun Yat-sen University, Guangzhou, China. His research interests include cultural studies, social movement studies, and communication studies. His articles have appeared in *Mass Communication Research, Asiascape: Digital Asia, Journal of Creative Communication, Chinese Journal of Communication*, and *International Journal of Communication*.

Shih-Diing Liu is an associate professor in the Department of Communication, University of Macau, Macao SAR. His articles have appeared in *Taiwan: A Radical Quarterly in Social Studies, Inter-Asia Cultural Studies, Third World Quarterly, Positions, Social Movement Studies, Interventions*, and *New Left Review*.

Iman Mitra is an assistant professor at the Centre for Development Practice and Research, Tata Institute of Social Sciences (TISS), Patna. He holds a PhD on the history of reception and popularization of the economic discipline in colonial Bengal at Jadavpur University and the Centre for Studies in Social Sciences, Calcutta (CSSSC). His research interests include history of economics, politics of translation, pedagogy and category formation in colonial and postcolonial South Asia, and urban history and migration. He has edited an anthology of essays titled *Accumulation in Postcolonial Capitalism* (Springer, 2016) with Ranabir Samaddar and Samita Sen.

Brett Neilson is a professor at the Institute for Culture and Society at Western Sydney University. With Sandro Mezzadra, he is the author of *Border as Method, or, the Multiplication of Labor* (2013). With Ned Rossiter, he has coordinated the project *Logistical Worlds: Infrastructure, Software, Labour*.

Ned Rossiter is professor of communication with a joint appointment in the Institute for Culture and Society and the School of Humanities and Communication Arts at Western Sydney University. His most recent book is *Software, Infrastructure, Labor: A Media Theory of Logistical Nightmares* (2016). With Brett Neilson, he has coordinated the project *Logistical Worlds: Infrastructure, Software, Labour*.

Ranabir Samaddar is Distinguished Chair in Migration and Forced Migration Studies at the Calcutta Research Group. His research focuses on migration and refugee studies, the theory and practices of dialogue, nationalism and postcolonial statehood in South Asia, and new regimes of technological restructuring and labour control. He is the author of many books and articles, most recently *Karl Marx and the Postcolonial Age* (Palgrave, 2017).

Kaustubh Mani Sengupta teaches history at Bankura University, India. He holds a PhD from the Centre for Historical Studies, Jawaharlal Nehru University, New Delhi. He was a postdoctoral fellow at the Transnational Research Group on 'Poverty and Education in India' funded by Max Weber Stiftung. His research focuses on urban history of South Asia, early colonial state in India, and history of infrastructure and space. His articles have appeared in journals such as *The Indian Economic and Social History Review, Studies in History,* and *South Asia Research*. He is working on a monograph, tentatively titled *Carving Calcutta: Space, Economy, and Law in Eighteenth-Century Bengal*.

List of Images

Image 2.1	Map of Calcutta from the *Imperial Gazetteer of India 1909*. Kidderpore is in the south where the river takes a westward turn. The warehouses and the *ghats* (landing places) dot the banks of the river on both sides, Calcutta on the east and Howrah on the west. (Image: Courtesy of the Digital South Asia Library, http://dsal.uchicago.edu)	26
Image 2.2	Balance scale in a warehouse near Howrah, 1944. (Photograph by Glenn S. Hensley, held by University of Chicago Library, Southern Asia Department)	38
Images 2.3 and 2.4	'Human Conveyor Belt'. (Photograph by Glenn S. Hensley, held by University of Chicago Library, Southern Asia Department)	42
Image 5.1	Woman throwing a chunk of coal down from a rail wagon that has stopped at a red signal, 2015. (Photograph by Samata Biswas)	106
Image 5.2	Left: Women scraping coal dust off coal-carrying trucks into cooking utensils as they wait at the level crossing, 2015. Right: Women carrying the gathered coal dust across NH41 in a bucket, 2015. (Photographs by Samata Biswas)	106
Image 5.3	Women and children digging in landfill material for iron scrap, 2015. (Photograph by Samata Biswas)	108

CHAPTER 1

Making Logistical Worlds

Brett Neilson, Ned Rossiter, and Ranabir Samaddar

Logistics makes worlds. What are the stakes of this claim? First, it indicates that logistics is productive. This may seem a truism for, at least in current dominant understandings, logistics is deeply implicated in capitalist production, where it is defined (in handbooks and management manuals) as the art and science of getting the right thing to the right place at the right time. Undoubtedly, the techniques and technologies that undergird logistics as a system of communication and transport have their productive sides because, in the current wave of globalization, the organization of supply chains and production networks has been a central feature: stretching out production across time and space, facilitating a constant movement of goods, people, and information across sites, trading labour costs against transport costs, and, in doing so, eroding the distinction between production and circulation. But to claim that logistics makes worlds is to say something more than that logistics makes commodities, supply chains, or even globalizing patterns of interconnection. At base it is in an ontological claim, and it is in this sense that we use it to explore the making of a world region.

B. Neilson (✉) • N. Rossiter
Institute for Culture and Society, Western Sydney University, Parramatta, NSW, Australia

R. Samaddar
Calcutta Research Group, Kolkata, West Bengal, India

By subtitling this volume 'The Labour of Making a World Region,' we focus our investigation around the Asian region. The chapters that follow evolved within a wider project, 'Logistical Worlds: Infrastructure, Software, Labour,' which began in 2011 as an investigation into the global expansion of Chinese interests through logistical and infrastructural installations. The project conducted research in and around three shipping ports in order to deepen our understanding of the processes underlying, and obstacles to, this expansion. The ports were Kolkata—an important and unavoidable choke point in plans to extend Chinese trade routes to the west, Piraeus—where the Chinese state-owned enterprise COSCO now operates the port, and Valparaíso—where plans to expand the port to receive large post-Panamax craft from China have been delayed.

Two of these sites are located beyond the Asian region as it is usually conceived—Piraeus is in Greece, ten kilometres from the city of Athens, and Valparaíso is on the Chilean coast, adjacent to the Santiago metropolitan region—but this was precisely the point. We sought to map how logistics stretches the cultural and geographical region of 'Asia' beyond its assumed boundaries and changes the configuration of, and relations between, its conventional subregions. An additional concern was to show that regionalism is partly constructed through global operations of infrastructure, software, and data rather than simply through culture, civilization, and geography. The implications of these operations for labour practices and their attendant modes of subjectivity offer a means of analyzing the relationship of logistical practices to capitalist crisis and transition. The current volume addresses these wider research concerns by focusing on the case of Kolkata port and drawing on a series of contributions that explore the relevance of logistics in China's rise as regional and global economic power.

China has a significant technical literature on logistics (物流) or *wùliú* in pinyin, meaning 'material flow' (Liu et al. 2016). Yet the role of China in the logistics industries has not been seriously examined in recent publications that have interrogated logistical practices, and modes of power, as a means of critically confronting contemporary capitalism (Toscano 2011; Bernes 2013; Cowen 2014). As an example, while Jesse LeCavalier's (2016) *The Rule of Logistics* is an extensive account of the logistical activities of Walmart in the US, it offers little detail on the company's operations in China. This is not to underestimate or dismiss recent contributions to the debate on logistics, to which the present authors have also contributed (Neilson and Rossiter 2011; Neilson 2012; Mezzadra and Neilson 2015;

Samaddar 2015; Rossiter 2016). All provide insights into the fluid positioning of states with respect to commercial enterprise, the remaking of urban and wider global spaces in response to the imperatives of growth and productivity, the role of software and data in the control of labour mobility, the interface of capitalist accumulation with processes of militarization and securitization, and the role of infrastructure as a scaffold for the computational, architectural, and technical organization of globalization.

Some contributors to this logistics debate have also occasionally focused their analytical gaze on Asian or Middle Eastern realities. Keller Easterling (2014, 25–69), for example, mentions developments in China's Shenzen, Korea's Songdo City, and Saudi Arabia's King Abdullah Economic City in her exploration of economic zoning practices. Deborah Cowen (2014, 163–195) presents Dubai Logistics City as a primary example in her analysis of the urban forms promoted by logistics enterprises. Benjamin Bratton (2015, 112–15) explores the conflict between China and Google as part of his totalizing geopolitical vision of the new forms of power wielded by what he calls ubiquitous planetary computing. None of these works, however, inquires systemically—which is to say, both historically and in relation to current capitalist formations—into the role of logistics in making world regions. Rather, in these studies, regions either tend to give way to an urban scale focus or appear more broadly as the effects of wider global dynamics.

The 'Logistical Worlds' project is not alone in examining how logistics makes regions. Laleh Khalili (2017) has carried out research on the ties between war and trade in the Middle East, based partly on fieldwork conducted on a container ship travelling between Malta and Jabal Ali in the United Arab Emirates. Similarly, Charmaine Chua (2015) has investigated what she calls the 'Chinese logistical sublime' by pursuing ethnography on a container vessel travelling from Tacoma in the US state of Washington to Yantian in China, and on to Taiwan. These studies have affinities with the work of this volume, that is, with the tracing of logistical routes, routines, and labour regimes in order to understand the changing internal dynamics and relations between world regions. This is especially important in the case of Asia, since much of the stretching of supply chains across global expanses that occurred with the so-called logistics revolution—which began in the 1970s and was given fillip by the 'opening' of China in the 1980s—was accompanied by the emergence of Asia, and particularly East Asia, as a favoured site of industrial production. These changes have implications for patterns of trade and investment, international divisions of labour, and the emergence of what Aihwa Ong (2006) calls 'lateral spaces,' that is, the combining of transcontinental production channels with racialized and gendered forms of labour segregation.

In recent times, the mode of globalization based on the transfer of production to Asia, by predominantly North Atlantic enterprises, has begun to shift. Programmes, such as China's Belt and Road Initiative, which aim to establish new trade routes from China—through Central and South Asia—to Europe and beyond, presage a new development in Asia-driven globalization that goes against the isolationist and protectionist predilections of the current US presidency. Announced by Xi Jinping in late 2013, and accompanied by the formation of the Asia Infrastructure Investment Bank, this initiative reconfigures China's geopolitical and geo-economic interests in relation to five major goals: policy coordination, facilities connectivity, unimpeded trade, financial integration, and people-to-people bonds (National Development and Reform Commission, Ministry of Foreign Affairs, and Ministry of Commerce of the People's Republic of China 2015). Alongside this development is India's emergence as a global economic power, the longer-standing role of countries such as Japan and South Korea as industrial powerhouses, and the strategic positioning of cities like Singapore and Hong Kong within global circuits of logistics and finance.

While the Belt and Road Initiative represents a major logistical expansion that has the potential to redesign global trade routes and financial strategies, scholarly commentary has for the most part approached this initiative within the classical idioms of international relations and comparative politics (Callaghan 2016; Ferdinand 2016; Yiping 2016), thus foregoing the opportunity to rethink processes of political and economic change through the analytical frame of logistics. Despite some exploration of the programme's financial implications (Sit et al. 2017) and the relevance of its cultural dimensions (Winter 2016), as well as the framing of its links to the decolonizing spirit of the 1955 Bandung conference (Wondam 2016), the Belt and Road Initiative's potential to stimulate debate on the intersection of logistics and capitalism has gone unremarked.

In using logistics to provide an epistemic angle for the analysis of contemporary capitalism and Asia-led globalization, however, the present volume does not just restrict its attention to the Belt and Road Initiative. A distinct challenge in broaching Asia, whether as research object or method (Chen 2010), is recognition of its indistinct boundaries and internal heterogeneity. This applies as much to the theoretical approaches that emerge in the human and social sciences as to questions of political organization, economic practice, social process, or cultural tendencies. Studies of China's rise, for instance, have generally drawn on approaches from world systems theory (see, for instance, Arrighi 2007 or Hamashita 2008) or interrogations of Chinese modernity (Wang 2011). By contrast, debates on India's transition have been informed by the uneasy intersection of Marxism with various paradigms of postcolonial thought. The current emphasis on 'postcolonial capitalism' in Indian discussions of economic and political transformation (see, for instance, Sanyal 2007; Mitra et al. 2016; Sinha 2016; Chatterjee 2017) draws attention to the combination of high financial and primitive accumulation, and the intermeshing of financial and labour mobility, in the rush towards logistical modes of organization that will supposedly generate spectacular growth and wealth. Crucial questions that are emerging in such debates take into account the dynamics of such logistical visions and growth; the fault lines in the accumulation regime—based on finance, extraction, and rent; the dynamics of migrant labour in relation to infrastructural expansion; and, finally, the impact of this logistical expansion on society, politics, and existing frameworks of sovereignty, legality, and gender. Yet the issue of their applicability across other sites in Asia remains open. Any attempt to export them is likely to be met with reminders of the specificity of Japanese debates about modernity (for a summary, see Walker 2016, chapter 2), for instance, or proclamations of how approaches to South-East Asian politics have influenced wider understandings of nationalism (Anderson 1983). Nonetheless, the debates surrounding postcolonial capitalism and the rise of China provide prominent and recognizable approaches to contemporary Asia—perhaps because of the positions of the nations with which they are associated—and the present volume uses the perspective of logistics to negotiate both, moving between China- and India-centred perspectives in ways that seek to relativize these styles of analysis and even interrogate the logistical connections that force us to think them together.

Perhaps it is due to the different histories of colonialism in China and India that analyses of their present roles in making inter-Asian connections draw on different—although related—intellectual strands. India passed in the mid-nineteenth century from a colonialism administered by the 'company state' of the English East India Company (Stern 2011) to direct administration by the British Crown. China, by contrast, was subjected to a form of concession colonialism following the Opium Wars, whereby parts of its territory were ceded to foreign powers through a series of treaties. A number of commentators (see, for instance, Nyíri 2009) have argued that China's colonial concessions provide a precedent for contemporary logistical territories established by it—both its internal Special Economic Zones and the spaces conceded to Chinese state enterprises in other parts of the world (an example of which is the concession of the port of Piraeus to China COSCO Shipping which is discussed by Nelli Kambouri and Pavlos Hatzopoulos in their contribution to this volume). Although we are wary of understanding current logistical developments as mere reversals of past imperial ventures, there is something to be gained from an analysis that emphasizes how the legal and geographical development of imperialism was shaped by commercial and logistical factors. Lauren Benton's study of the making of legal infrastructures of sovereignty and jurisdiction in early modern imperialism shows how European empires tended to exercise control 'mainly over narrow bands, or corridors, of territory and over enclaves of various sizes and situations' (Benton 2005, 700). The fact that this kind of fragmented territorial control took a different path in India than in China suggests much about the forms of colonialism that emerged in these continental spaces. By the time of the Raj, the British Crown administered a vast territory on the subcontinent and instituted many legal measures that have influenced the forms and practices of capitalism in the postcolonial era: for instance, the Land Acquisition Act of 1894, which continued to apply to practices of governmental land expropriation until 2014, was often used in combination with the Special Economic Zones Act of 2005 to establish new kinds of logistical and industrial spaces. Such land acquisition has been central to the debates about postcolonial capitalism that have animated political and cultural discussion in India in recent years. In considering the differences that invest China- and India-centred perspectives on the changing forms and faces of a region called Asia, it is thus essential to keep the past in view.

The chapters that follow certainly do not shy away from this historical burden. Equally, in putting a logistical gaze upon contemporary capitalist dynamics, they do not adopt an approach that emphasizes local or national specificities to the point that they obscure the commonalities of capitalism across diverse contexts. In this way, the volume seeks to avoid the pitfalls of a perspective that finds global capitalism to structure all differences while also questioning a position that stresses particularity to the point where the unevenness of capitalism overshadows its systemic features (see Murthy and Liu (2017) for a discussion of how this antimony plagues non-Western trajectories of Marxism). By focusing on how logistical operations play out on the ground, we seek to understand how capital intervenes in specific sites and situations to shift and redesign social relations in its image. At the same time, we are careful not to reduce capitalism to logistics, remaining attentive to how logistical operations concatenate with each other as well as with other operations of capital. In this regard, the relation of logistics to finance is crucial.

If finance provides the abstract point of coordination for contemporary capitalism, logistics provides the material nexus of its coordination. In addition to the mutual implication of logistics and finance, however, there is the role of extraction—both of natural resources and patterns of social cooperation. Together with other operations of capital, finance, logistics, and extraction compose the unstable whole of capitalism (Mezzadra and Neilson 2015). Yet this process of composition, which is also one of concatenation, does not mean that capitalism can be understood as an accomplished totality. By exploring how capital's operations prospect and draw upon its multiple outsides, we observe how it is driven by a logic of accumulation and unlimited expansion while, at the same time, tracking how its operations bring particularity into relationship with universality, and how the complexities of this interrelationship become apparent across diverse Asian contexts. More specifically, in asking how logistical operations of capital create routes, connections, and spaces within, across, and beyond the region, we understand that they not only enact commercial imperatives but also embody novel forms of political power. An important part of the study, therefore, is to understand how these forms of power relate to those that continue to be embodied in and expressed by the state.

Port as Infrastructure of Postcolonial Capitalism

The volume is divided into two parts. The first is titled 'Port as Infrastructure of Postcolonial Capitalism.' Building on the discussions of postcolonial capitalism outlined above, it presents six studies that examine the historical and contemporary logistical complexities surrounding the port of Kolkata. We have chosen this port as a site of in-depth investigation because it has long been an unavoidable choke point in the making of inter-Asian connections as well as Asia's connections with the world at large. In the transitions from the historical triangular trade between China, India, and Britain to the current Chinese Belt and Road Initiative, Kolkata port's fortunes have fallen repeatedly only to rise again despite the challenges posed by its location as a river port. While our conceptual understanding of logistics extends beyond transport industries, shipping ports remain crucial and iconic facilities in the organization of patterns of production and trade. They are sites of transit and calculation where the intermodal logic of containerized transport meets technologies of digital control and the fractious politics of labour. The investigation of Kolkata port provides an opportunity to track these changes and conflicts from colonial to postcolonial times, and examine how logistics intersects the transformations of capitalism and the making of wider regional connections.

In the background of the geopolitical and economic considerations that facilitate Chinese investment in ports in Pakistan, Bangladesh, and Sri Lanka, a study of the historical, financial, and spatial transformations surrounding the port of Kolkata offers a means of tracking and analyzing the frictions and tensions that inevitably accompany, and are produced by, logistical efforts to create smooth spaces for the circulation of people, things, and capital. The fact that most Asian ports run terminal operating systems manufactured by the South Korean company, Total Soft Bank, while North American and European ports run software developed by the California-based firm, Navis, raises issues of technical interoperability that parallel but cannot be reduced to questions of cultural translation. This, in turn, has implications for data or process mining techniques deployed in those ports, with knock-on effects for labour regimes, human-machine interactions, and the passage of goods. Practices of transshipment, which involve the transfer of containers from one ship to another without unloading or customs inspections, shift the relations between regions as traditionally defined. For instance, the fact that almost every container that arrives in Kolkata is transshipped through Singapore places those cities in what might be called the same logistical region, tied not only by shipping routes but also by regulatory regimes. Yet, within the received denominations of area studies, the former is in South Asia and the latter in South-East Asia. The study of logistics in this instance allows new analytical questions to be asked about the changing stakes of designating Asia as a region, but such questions are blocked and indeed remain unanswerable within the bounded analytical areas and disciplinary horizons of post-war academic knowledge.

A study of economic relations and material conditions also enables us to investigate what can be called the spectral presence of labour in logistical sites such as ports. In the case of Kolkata, this presence is particularly marked as much of the labour—for instance, that employed in unloading ships at the Kidderpore docks—is performed by migrant workers from other parts of India who, contrary to the standards of measure introduced by containerization, carry sacks on their backs and are paid according to weight moved rather than volume transported. At the same time, the port supports much labour not directly involved in the movement of freight, including workers in the highly informal economy that surrounds and supports the operations of the port, including those related to corruption and crime. All this—along with questions of gender, unionization, finance, urban planning, and dispossession—is taken up in the in-depth exploration of the logistical operations of Kolkata port undertaken in this part of the book.

The chapter by Kaustabh Mani Sengupta analyzes the development of the port facilities in Kolkata from the third quarter of the nineteenth century. It specifically examines the way goods were brought to the port, stored there, and then shipped or transported to other areas. The modes and mechanism of the port facilities are studied against the general political and economic backdrop of the times. Industrial growth in India and massive increases in import/export trade necessitated rapid development of port infrastructure in Kolkata. The Calcutta Port Trust was officially established in 1870 and made rapid advances in building additional jetties and streamlining dock logistics and cargo handling. Sengupta focuses on two important aspects of this enterprise. Firstly, the role of warehouses in facilitating the trading activities of the port, the negotiations that took place amongst the various actors in constructing these facilities, and the problems faced in maintaining them. And secondly, the crucial part played by the transport system in aiding the movement of goods to and from the port area. Both enterprises reveal how a contestation of territory ensued, the way various interest groups operated, and the effect of political-economic considerations on shaping the city's river-front space. The chapter also notes the contingencies in port planning, the measures adopted for safety and security, and the alterations or deviations in shaping the port infrastructure. Finally, a close reading of the modes and mechanisms of construction of the port over time offers a glimpse into how everyday logistics establishes a complex of men, machine, and things.

Iman Mitra seeks to understand the entangled framework of infrastructure, software, and labour from the specific yet interconnected perspectives of the spatialization of calculability and the financialization of space. He approaches Kolkata port as a site where these two perspectives collide and give rise to a particular form of logistical governance.

This form of governance requires negotiations with, and navigations through, a network of institutional apparatuses which produce the material basis of calculations and speculations that envisage the material connection between infrastructure, software, and labour. Kolkata Port Trust is one such institutional apparatus. Founded in 1870 by the colonial rulers in India, it was bestowed with the responsibility for the expansion and management of the Calcutta Port, which included the carrying out of endless calculations and speculations related to the port's geopolitical exclusivity. Mitra shows how correspondence between navigational calculations and speculations regarding space-making exercises (including rent extraction from the land owned by the Port Trust in the city) gives birth to a vision of logistics that involves various stakeholders in the processes of global capitalist expansion against the backdrop of the growing recognition of the port's locational advantage in schemes such as India's Look East Policy or China's Belt and Road Initiative.

No discourse on logistics is complete without a discussion of crime. The routes through which goods and ideas move are often the same routes used by networks and organized crime groups. Legitimate business hubs often coexist alongside markets for smuggled goods. Sometimes the players are one and the same. In Paula Banerjee's exploration of the nexus of crime and logistics, in and around the Kidderpore docks in Kolkata, she considers the gendered and social dimensions of this situation, both historically and in the present day. Banerjee examines the logistics of port crimes and how they transform the periphery into a central question of enquiry. A port is often at the periphery of an urban logistical system, but sometimes it can transform itself into the main logistical hub for the development of urban space. When that happens, crimes in the port also become central to the security issues of the city, and Kolkata is no exception in this regard. Banerjee thus asks why crimes that happen in the port have much larger ramifications. Even when these crimes may appear random, there is always some logic behind why they happened in a particular logistical space. Anything happening in the port discursively spreads like wildfire. One reason perhaps is that a port cannot be contained as its main function is dispersal. Also its contacts with the outside world make it a problematic space from the perspective of security because, in the administrative imagination, all that is threatening to stability comes from the outside. Studying how the state attempts to tightly control a space that is allocated for all kinds of movement raises the question of the relation of logistics to political order, in terms of both how movement is regulated and how logistics generates its own forms of political power.

Samata Biswas' chapter explores the logistical space of Haldia, where the deep-sea dock complex of Kolkata Port Trust is located, to analyze how a port city is created by logistical imperatives and the way it shapes new forms of logistics. Located near the mouth of the river Hooghly, Haldia is a city, a municipality, a riverine port, and an industrial belt. Any attempt to map the logistical networks that cross Haldia—in the form of warehouses, pipelines, roads, trucks, local transport, container traffic, human beings, and cattle—has to pay attention to the interconnections between industry, the dock complex, the geographical area, the human actors, and their nonhuman interests. Based on ten months of ethnographic research, with an emphasis on the collection of visual data, Biswas analyzes five figures that stand in for the logistical transformations underway in Haldia: the illiterate owner of a large logistics firm, the recently closed Renuka Sugar Mill, a local woman who collects coal dust along the road to the port in order to make and sell pellets, the Ural India factory, and the absent figure of public health. Through a close examination, as well as of the subjectivities associated with these figures, Biswas details the existence and flourishing of an 'other' logistics in Haldia, one based in modes of survival and hustling that exist in the shadows of the city's official logistical activities. This approach allows analysis of the changing relations between formal and informal labour, the emergence of new forms of extraction, and the relevance of diverse scales, spaces, and histories in current formations of postcolonial capitalism.

The changing geopolitics of Asia, marked by China's sharp rise and India's emergence, have led to a renewed importance for the Kolkata port system, even as questions have been raised about its future due to poor draught (low water depth), age-old infrastructure, and the high cost of operations. Indian policy makers may worry about a possible Chinese maritime encirclement by a 'string of pearls' (including the China-constructed ports of Gwadar in Pakistan, Hambantota in Sri Lanka, and Kyauk Pyu in Myanmar), but the Kolkata Port Trust regards a China-India road and a reaching out to Tibet as possible options to augment the future business of the port. The Chinese, for their part, have identified the Kunming-Kolkata (K2K) corridor, now known as part of the proposed BCIM (Bangladesh-China-India-Myanmar) corridor, as one of the six economic corridors to be developed under President Xi Jinping's Belt and Road

Initiative. Subir Bhaumik's chapter considers five infrastructural projects for the improvement of Kolkata's port system, including the construction of a deep-sea facility at Sagar Island, and relates them to wider logistical and geopolitical initiatives that are reshaping the Asian region. Bhaumik also examines the relevance of the Indian government's identification of Kolkata as the starting point of its 'Act East' thrust. In short, Bhaumik approaches Kolkata port as a vital switch point for contemporary infrastructural projects that seek to remake Asian regionalism by connecting India to its neighbours in the East and eventually to China.

LOGISTICS OF ASIA-LED GLOBALIZATION

The second part of the book is called 'Logistics of Asia-led Globalization.' Unlike the first part, which focuses on a single logistical facility, the second part's seven chapters investigate how diverse sites and modalities of logistics rearrange relations within and between Asia's subregions and remake Asia's relations with the world. The topics here are diverse: ranging from a gateway in North Bengal in India—where a dusty transit town connects the Indian mainland, Bangladesh, Nepal, Bhutan, and India's Northeast, and determines the nature of economic corridors in that subregion—to Chinese investment in the Greek port of Piraeus, the politics of e-waste recycling in Hong Kong, the politics of economic corridors, and finally the labour of Internet use in China. All of these interventions show how logistics generates dependencies, conflicts, unwanted circulations, and political orders, and how the latter in turn produces particular forms of circulation. This link between logistical circulation and political orders becomes especially evident in consideration of the new kinds of territory produced by logistical facilities. Understanding the varieties of political order, power, and space that such installations generate means documenting how logistical practices can reconfigure territory in ways that rival and parallel the traditional territoriality of the nation-state.

Atig Ghosh's chapter acts as a bridge between the book's two parts by extending the geographical scope of analysis to examine the logistical transformations surrounding the North Bengal city of Siliguri, a key site of interchange in the hinterlands of Kolkata port. When the partition of India created a geographical barrier in the north-eastern part of the country, the narrow Siliguri Corridor—commonly known as the Chicken's Neck—remained as a national-territorial isthmus between the north-eastern part of India and the rest of the country. Siliguri thus found itself pitchforked to a position of immense geostrategic importance. Wedged between Bangladesh to the south and west and China to the north, Siliguri has no access to the sea closer than Kolkata, on the other side of the corridor. Ghosh's chapter investigates the importance of logistics to the town's fairy tale growth, emphasizing the role of different actors and infrastructures. A site of military build-up and territorial anxieties, Siliguri has traditionally survived on the basis of four industries: timber, tea, tourism, and transport. To these it is necessary to add a fifth T, that is, the trafficking of humans, principally young women for the sex trade. Ghosh investigates how the logistical convergence of these industries creates a political-economic *mélange* of people in flux—wholesalers, retailers, traders, military and security personnel, tea planters, trafficked bodies and their consumers, gun-runners, political fugitives, asylum-seekers, railway men, construction workers, and stateless groups. Special attention is given to the transitory or 'floating' character of Siliguri's economy—the way it hinges on fluxes of migration and land grabs as well as the transformative construction of the Asian Highway, which promises to open new paths for capital to South-East Asia and China.

In 2009, labour relations in the port of Piraeus were radically transformed as a result of a concession agreement signed between the Greek government and COSCO Pacific Ltd. Studying the transformations of labour and logistics at Piraeus, Nelli Kambouri and Pavlos Hatzopoulos offer insights into changing patterns of human struggle and precarity in the wake of the port's emergence as a key station within China's Belt and Road Initiative. Kambouri and Hatzopoulos draw on research fieldwork begun in 2013 for an analysis of precarious labour in regimes of logistical governance as these were shaped in the port of Piraeus following the COSCO concession. For this analysis, they employ the concept of the machinic assemblage which denotes, in the context of the chapter, entanglements of machines, humans, software, and discourses that produce relations of power exercised through the logic of control. Along these lines, they approach operations in the Piraeus container terminal as a social machine. The container terminal takes human labour in its gears, along with containers and all the machinery required to move them around the terminal: quay cranes, rail-mounted and rubber-tyred gantry cranes, as well as the trucks and software platforms that generate and control their complex movements through algorithmic computations. From this perspective, they argue that labour in the Piraeus container terminal can no longer be represented by the image of workers using their bodies to perform repetitive tasks in order to operationalize machines. Labour in Piraeus is enabled, instead, through the everyday functioning of cybernetic organisms whose lives are ordered according to the objectives of maximized efficiency and minimalized idleness. Investigation of these transformations allows Kambouri and Hatzopoulos to undertake a critical analysis of the gendered and racialized dimensions of accounts that seek to explain changes to the labour regime in Piraeus, whether seen as a 'Chinification' of Greece or the establishment of a Chinese economic zone in the European Union.

Corridors are an emerging catchall in logistical discourses. From trade and investment corridors to freight corridors, from digital corridors to development corridors, from transport corridors to industrial corridors, it is hard to avoid the reference to this concept in discussions of logistics. Giorgio Grappi's analysis examines how corridors are being used as both an organizational tool and a political concept that marks the language of logistical governance. While the *supply chain* refers to the pervasive economic process that underlies logistics, the language of *corridors* refers to the materiality of infrastructure—that makes possible logistical operations at a larger scale—and, most importantly, to the so-called soft infrastructure of governance. Corridors impose a diverse territoriality and modify the functioning of existing institutions by imposing technical standards, governance tools, and financial flows, as well as producing a variegated geography of logistical power. Introducing the concept of the *politics of corridors*, and addressing examples from China's Belt and Road Initiative and the European Union's TEN-T project, Grappi investigates how languages and techniques developed in policy papers, master plans, and international studies are producing a new political discourse and what this suggests and implies in relation to power, the politics of the state, and political theory.

Rolien Hoyng questions the appearance of communication and the knowledge economy in ways that render infrastructure and hardware invisible and make us forget about their material support. Approaching Hong Kong as what she calls a 'dirty' smart city, Hoyng focuses on e-waste recycling and the socio-material relations this constitutes. She analyzes the management of the mobility and materiality of e-waste in the formal recycling industry and explores its relation to the informal sector. Hoyng rethinks the binary of order/disorder, which underlies much of the critical literature on waste, by connecting it to questions of power and governance.

Furthermore, she asks to what extent the informal sector of e-waste recycling either undermines or complements operations of the formal knowledge economy. The chapter focuses on the analysis of legal regulations, licences, permits, certificates, and software—as well as waste—as transient matter that manages to generate *and* subvert socio-material relations. Her investigation of the nexus of governance and materiality in e-waste recycling in Hong Kong allows an analysis of logistical practices and relations that actively remake and rechannel regional flows in a key Asian city.

Taking the China-Pakistan Economic Corridor (CPEC) as its point of empirical engagement, Majed Akhter's chapter sketches a theoretical orientation to the geopolitical analysis of 'global China' and China-in-Asia. Akhter understands China's Belt and Road Initiative in the context of an economic crisis of over-accumulation and points to a series of 'spatial fixes' that China has undertaken to resolve this situation through the creation of an infrastructurally integrated Asia with China at the centre. In this way, he charts a path beyond state-centred visions of the Belt and Road Initiative while also arguing that large infrastructures are projects not only of accumulation but also of state territorialization. Studying deep-rooted socialhistorical structures that have led to the militarization of logistical corridors and enclaves in Pakistan, Akhter examines how the state *secures* the infrastructural operations of capital. This analysis leads to a reading of the CPEC that understands it not as a means of creating a smooth space for transnational flows of capital and commodities but as a reassertion of deeply heterogeneous and fragmented social space as produced over generations in Pakistan's peripheries. Akhter thus seeks at once to contextualize the politics of large infrastructural projects within the contradictions of capital accumulation on a world scale and to show how China's Asiacentric vision of globalization runs up against spatial and social realities that it can neither completely homogenize nor control.

Zhongxuan Lin and Shih Ding Liu shift the debate on logistical infrastructure away from ports, corridors, and territoriality by arguing that, under current capitalist conditions, the Internet, media, and software function as vital infrastructures for the production and processing of information. Working from Macau—a historical junction of the coolie trade, the only legal site for gambling in China, and a major importer of labour from the mainland and South-East Asia—they question the tenets of the debate on immaterial labour (or labour that does not produce a material good or product) as conceived by European—especially Italian—thinkers.

They ask what Internet users in Macau produce, how they produce, and why they produce. Lin and Liu thus bring a widely different perspective on the production of the Asian region through logistics by emphasizing the role played by collective sentiment and predisposed social relationships, as well as the affective motivations that propel and sustain such labour. In seeking to dislodge some of the most powerful theoretical frameworks for understanding the subjectivity of immaterial labour, they do not take the specificity of Internet use in Macau as a definitive ground for theory building or concept production. Rather they call for further research and comparison to investigate the involvement of digital communication systems in the making of world regions. In this way, Lin and Liu offer an analysis of Asia-led globalization that is less about the flows and connections that enable the circulation of capital and information, within and beyond the region, than the ways in which the production of space is necessarily linked to the production of subjectivity.

A final chapter by Brett Neilson explores the productivity and limits of a methodological principle—*follow the software*—that has informed the 'Logistical Worlds' project in conducting research into the interactions of labour, software, and infrastructure in three shipping ports: Piraeus, Kolkata, and Valparaíso. Rather than following the money or following commodities, Logistical Worlds has addressed empirical investigations to sites where the interoperativity between software systems breaks down. The hunch has been that these sites will also reveal political conflicts, social inequalities, cultural transactions, labour struggles, and infrastructural workarounds that are constitutive for the logistical project of making things circulate. The chapter pays special attention to the spatial qualities of these sites, arguing that they give rise to forms of territory and territoriality that rival and parallel the forms of political and spatial order established by the state. Tying together the other chapters in the section, Neilson suggests that Logistical Worlds has enabled a rethinking of the deep processes of heterogenization that remake political spaces in their tense entanglement with spaces of capital.

Coda

Taken together, the two sections of the book examine logistical practices and infrastructural activities that reshape Asia's internal dynamics and external boundaries. The chapters in the volume collectively explore how logistical practices of planning and organization interact with historical and contemporary social realities, including the changing global positions of China and India. In this nexus between logistics, society, and history, it becomes possible to understand the political imperatives within which logistics functions as well as the organization of dissent, resistance, and violence that it often occasions. Logistics affects the distribution of waste and resources, the livelihood and dispossession of workforces, the gendered dimensions of social reproduction, and, not least, the violence, crime, and parallel world of subaltern globalization. Logistics spans and confuses the distinction between formal and informal sectors while also structuring life in adaptive ways that constantly shift in response to the environment and feed back into prevailing material conditions.

The ability of logistics to adjust to contingencies says much about its pervasive power. It also explains why logistics is so difficult to pin down. Studies in this field tend to quickly shift their frames of reference. One moment logistics is a socio-technical system, then it is a set of organizational modes, a network of spaces, a corridor for commodities, and so on. This collection seeks to cut through this multiplicity of perspectives by asking how logistics provides an unexamined background for contemporary ways of being, knowing, and living. The question of political order, power, and space emerges strongly in those chapters dealing with the making of logistical corridors and spaces. While research on export processing and other kinds of economic zones has dominated the geographical investigation of logistics, the volume also reflects a growing attention to the making of large-scale spaces of transit, communication, and production that reach 'beyond zoning.' This work reflects and builds upon an earlier concern with large-scale infrastructural installations linked to, say, irrigation, transport, or pipelines, but it also responds to the new realities of accumulation through logistics—a feature of the capitalism of our time.

In sum, the volume is perched on an interrogation of how logistics contributes to the making and unmaking of the world region called Asia and, in turn, how this making and unmaking of the Asian region shapes global logistical practices. The book also investigates how logistics generates dependencies, conflicts, circulations, and political orders, and how, in turn, political orders produce certain forms of circulation. The overall inquiry thus has implications for rethinking relations between capital, empire, and state as well as for questions of social governance and the production of subjectivity.

References

Anderson, Benedict. 1983. *Imagined Communities: Reflections of the Origins and Spread of Nationalism*. London: Verso.
Arrighi, Giovanni. 2007. *Adam Smith in Beijing: Lineages of the Twenty-First Century*. London: Verso.
Benton, Lauren. 2005. 'Legal Spaces of Empire: Piracy and the Origins of Ocean Regionalism.' *Comparative Studies in Society and History* 47, no. 4: 700–24.
Bernes, Jasper. 2013. 'Logistics, Counterlogistics, and the Communist Prospect.' *Endnotes* 3. Accessed 10 April 2017. https://endnotes.org.uk/issues/3/en/jasper-bernes-logistics-counterlogistics-and-the-communist-prospect
Bratton, Benjamin. 2015. *The Stack: On Software and Sovereignty*. Cambridge, Mass.: The MIT Press.
Callaghan, William A. 2016. 'China's "Asia Dream": The Belt Road Initiative and the New Regional Order.' *Asian Journal of Comparative Politics* 1, no. 3: 226–43.
Chatterjee, Partha. 2017. 'Land and the Political Management of Primitive Accumulation.' In *The Land Question in India: State, Dispossession, and Capitalist Transition*, edited by Anthony P. D'Costa and Achin Chakraborty, 1–15. Oxford: Oxford University Press.
Chen, Kuan-Hsing. 2010. *Asia as Method: Toward Deimperialization*. Durham: Duke University Press.
Chua, Charmaine. 2015. 'Slow Boat to China.' *The Disorder of Things* (blog). 5 January–7 February. https://thedisorderofthings.com/tag/slow-boat-to-China/
Cowen, Deborah. 2014. *The Deadly Life of Logistics: Mapping Violence in Global Trade*. Minneapolis: University of Minnesota Press.
Easterling, Keller. 2014. *Extrastatecraft: The Power of Infrastructure Space*. New York: Verso.
Ferdinand, Peter. 2016. 'Westward Ho – The China Dean and "One Belt, One Road": Chinese Foreign Policy under Xi Jinping.' *International Affairs* 92, no. 4: 941–57.
Hamashita, Takeshi. 2008. *China, East Asia and the Global Economy: Regional and Historical Perspectives*. Abingdon: Routledge.
Khalili, Laleh. 2017. *The Gamming* (blog). 22 February–1 March. https://thegamming.org/
LeCavalier, Jesse. 2016. *The Rule of Logistics: Walmart and the Architecture of Fulfillment*. Minneapolis: University of Minnesota Press.
Liu, Biang-lian, Ling Wang, Shao-ju Lee, Jun Liu, Fan Qin, and Zhi-lin Jiao, eds. 2016. *Contemporary Logistics in China: Proliferation and Internationalization*. Heidelberg: Springer.
Mezzadra, Sandro and Brett Neilson. 2015. 'Operations of Capital.' *The South Atlantic Quarterly* 114, no. 1: 1–9.

Mitra, Iman Kumar, Ranabir Samaddar and Samita Sen, eds. 2016. *Accumulation in Post-Colonial Capitalism*. New Delhi: Springer.
Murthy, Viren and Joyce C.H. Liu. 2017. 'Introduction: Marxism, Space, Time and East Asia.' In *East Asian Marxisms and Their Trajectories*, edited by Joyce C. H. Liu and Viren Murthy, 1–10. Abingdon: Routledge.
National Development and Reform Commission, *Ministry of Foreign Affairs, and Ministry of Commerce of the People's Republic of China*. 2015. *Vision and Actions on Jointly Building Silk Road Economic Belt and 21st-Century Maritime Silk Road*. March. Accessed 10 April 2017. http://en.ndrc.gov.cn/newsrelease/201503/t20150330_669367.html
Neilson, Brett. 2012. 'Five Theses on Understanding Logistics as Power.' *Distinktion: Scandinavian Journal of Social Theory* 13, no. 3: 323–40.
Neilson, Brett and Ned Rossiter. 2011. 'Still Waiting, Still Moving: On Labour, Logistics and Maritime Industries.' In *Stillness in a Mobile World*, edited by David Bissell and Gillian Fuller, 51–68. Abingdon: Routledge.
Nyíri, Pál. 2009. 'Foreign Concessions: The Past and Future of a Shared Form of Sovereignty.' *Espaces Temps*, 23 November. Accessed 17 April 2017. https://www.espacestemps.net/articles/extraterritoriality-pal-nyiri/
Ong, Aihwa. 2006. *Neoliberalism as Exception: Mutations in Citizenship and Sovereignty*. Durham: Duke University Press.
Rossiter, Ned. 2016. *Software, Infrastructure, Labor: A Media Theory of Logistical Nightmares*. New York: Routledge.
Samaddar, Ranabir. 2015. 'Zones, Corridors, and Postcolonial Capitalism.' *Postcolonial Studies* 18, no. 2: 208–21.
Sanyal, Kalyan. 2007. *Rethinking Capitalist Development: Primitive Accumulation, Governmentality and Post-Colonial Capitalism*. New Delhi: Routledge.
Sinha, Subir. 2016. '"Histories of Power", the "Universalization of Capital", and India's Modi Moment: Between and Beyond Postcolonial Theory and Marxism.' *Critical Sociology*, 43, nos. 4–5: 1–16.
Sit, Tsui, Erebus Wong, Lau Kin Chi, and Wen Tiejun. 2017. 'One Belt, One Road: China's Strategy for a New Global Financial Order.' *The Monthly Review* 68, no. 8: 36–45.
Stern, Philip. 2011. *The Company State: Corporate Sovereignty and the Early Modern Foundations of British Empire in India*. Oxford: Oxford University Press.
Toscano, Alberto. 2011. 'Logistics and Opposition.' *Mute*. 9 August. Accessed 10 April 2017. http://www.metamute.org/editorial/articles/logistics-and-opposition
Walker, Gavin. 2016. *The Sublime Perversion of Capital: Marxist Theory and the Politics of History in Modern Japan*. Durham: Duke University Press.
Wang Hui. 2011. *The Politics of Imagining Asia*. Cambridge, Mass.: Harvard University Press.

Winter, Tim. 2016. 'One Belt, One Road, One Heritage: Cultural Diplomacy and the Silk Road.' *The Diplomat.* 29 March. Accessed 10 April 2017. http://thediplomat.com/2016/03/one-belt-one-road-one-heritage-cultural-diplomacy-and-the-silk-road/

Wondam, Paik. 2016. 'The 60th Anniversary of the Bandung Conference and Asia.' *Inter-Asia Cultural Studies* 17, no. 1: 148–57.

Yiping, Huang. 2016. 'Understanding China's Belt & Road Initiative: Motivation, Framework and Assessment.' *China Economic Review* 40: 314–21.

PART I

Port as Infrastructure of Postcolonial Capitalism

CHAPTER 2

The Port of Calcutta in the Imperial Network of South and South-East Asia, 1870s–1950s

Kaustubh Mani Sengupta

The row of four massive warehouses standing along the Hooghly riverfront on Strand Road, once statements of the city's power and prosperity, had over the years become symbols of utter neglect on the part of the Calcutta Port Trust, that owns them, and the city fathers as well.

Strand Warehouse, the skeleton of which stands at the crossing of Brabourne Road, was the oldest, most ornate and aesthetically pleasing of the four. It was pushed to dereliction by several fires, the last and most devastating of which was on February 14. Deeper north, there are several other warehouses once owned by Bengali merchants, the most picturesque of which is the celebrated but disintegrating Putul Bari overlooking Sovabazar jetty.

These four warehouses were constructed between 1901 and 1903, Calcutta's boom time, and were the city's moorings on the Hooghly. They were the gateway to the city for shippers and when they left, their vessels used to be loaded with shellac, linseed, tea and gunnies. (*The Telegraph*, 30 May 2010)

Notwithstanding this picture of doom and despair, the port at Calcutta is still functioning and remains an important node in the circuit of trade between the subcontinent and East and South-East Asia. Nevertheless,

K. M. Sengupta (✉)
Department of History, Bankura University, Bankura, India

© The Author(s) 2018
B. Neilson et al. (eds.), *Logistical Asia*,
https://doi.org/10.1007/978-981-10-8333-4_2

there have been changes over the years in terms of its functioning as the port and in its relationship with the rest of the city. The celebrated warehouses on the Strand Road—the promenade on the eastern side of Hooghly, separating the city from the river—were once grand structures with an important mercantile function and they played a crucial role in augmenting the business of the port at Calcutta. In the imperial map of Britain, Calcutta held a central place in organizing the trade of the empire. The port—with its jetties, warehouses, wharves, railway, and tramway connections—facilitated maritime trade as well as overland trading activities.

From the latter part of the nineteenth century, industrial growth in India and a massive increase in import/export trade necessitated rapid development of port facilities in Calcutta. While sharing the major percentage of export trade in India during the early half of the twentieth century, the Calcutta port also emerged as a crucial entry-point for goods destined for the eastern and northern provinces of the subcontinent. Excellent transport routes, especially railways, acted as a catalyst for trading activities and, despite its unfavourable geographical features, the Calcutta port became the leading centre of trade and commerce for the colonial state in India. After the Calcutta Port Trust was officially established in 1870, it made rapid progress in building additional jetties and streamlining dock logistics and cargo handling.

In this chapter, I look at the development of the port facilities in Calcutta from the third quarter of the nineteenth century and examine the modes and mechanisms of the port facilities—the way goods were brought to the port, stored there, and then shipped or transported to other areas. Against the general political and economic backdrop of the times, I focus on two important aspects of the port's enterprise. Firstly, the role of warehouses in facilitating the trading activities of the port, the negotiations that took place among the various actors in constructing these places, and the problems faced in maintaining them. And, secondly, the crucial part played by the transport system in aiding the movement of goods to and from the port area. Both enterprises reveal how territory became contested, how various interest groups operated, and how political-economic considerations shaped the space of the city along the river front. I also note the contingencies adopted in planning for the port, the measures taken for safety and security, and the alterations or deviations in shaping the infrastructure.

Such a close reading of the modes and mechanisms of construction around the port provides a glimpse into the everyday logistics of establishing this complex of men, machines, and things. Humans appropriate natural space to build new spatial elements—such as roads and canals, villages, towns, and markets—and, following Henri Lefebvre, as Ravi Ahuja argues:

> These spatial elements are not simply 'things' – they are *at once* locatable objects and spatial relations. Social spaces are constituted through a complex of such relations – spatial relations that are inseparably integrated with relations between social groups, with property relations in general and relations of land control in particular. (Ahuja 2009, 25)

The individual stories of the warehouses and the transport around the port reveal these negotiations and relations. But before going into the details of these activities, it is important to briefly outline the larger political and economic context for this chapter (Image 2.1).

A Short History of Calcutta Port

S. C. Stuart-Williams, Vice-Chairman and then Chairman of the Calcutta Port Commissioners in the 1920s, delineated the area under the Calcutta port system to an audience in London in the following manner:

> The jurisdiction of the Calcutta Port Commissioners is of two kinds, namely, that within the port proper, which now commences at Konnangar, eight miles above Calcutta, and terminates at the subsidiary oil port of Budge Budge, thirteen miles below Calcutta, and also the more limited jurisdiction over the headwaters of the river and Port Approaches, the former of which commences at Kalna, seventy miles above Calcutta, terminating at the upstream limit of the port proper, and the latter commencing at Budge Budge, and terminating at the Sandheads. The whole of their jurisdiction thus comprises nearly 200 miles of river proper, its headwaters and the estuary. (Stuart-Williams 1928, 891)

Stuart-Williams identifies a new phase in the development of the Calcutta port complex that took place from the 1880s. This was facilitated by the increase in import trade which brought about a corresponding increase in export trade as well as a greater demand for facilities to accommodate the steamers that were replacing sailing vessels. In this time, the capacity of the docks was stretched to 27 berths, of which 17 were devoted to the export trade. As Stuart-Williams notes, 'In this period the accommodation available may be said to have been definitely overtaken by

26 K. M. SENGUPTA

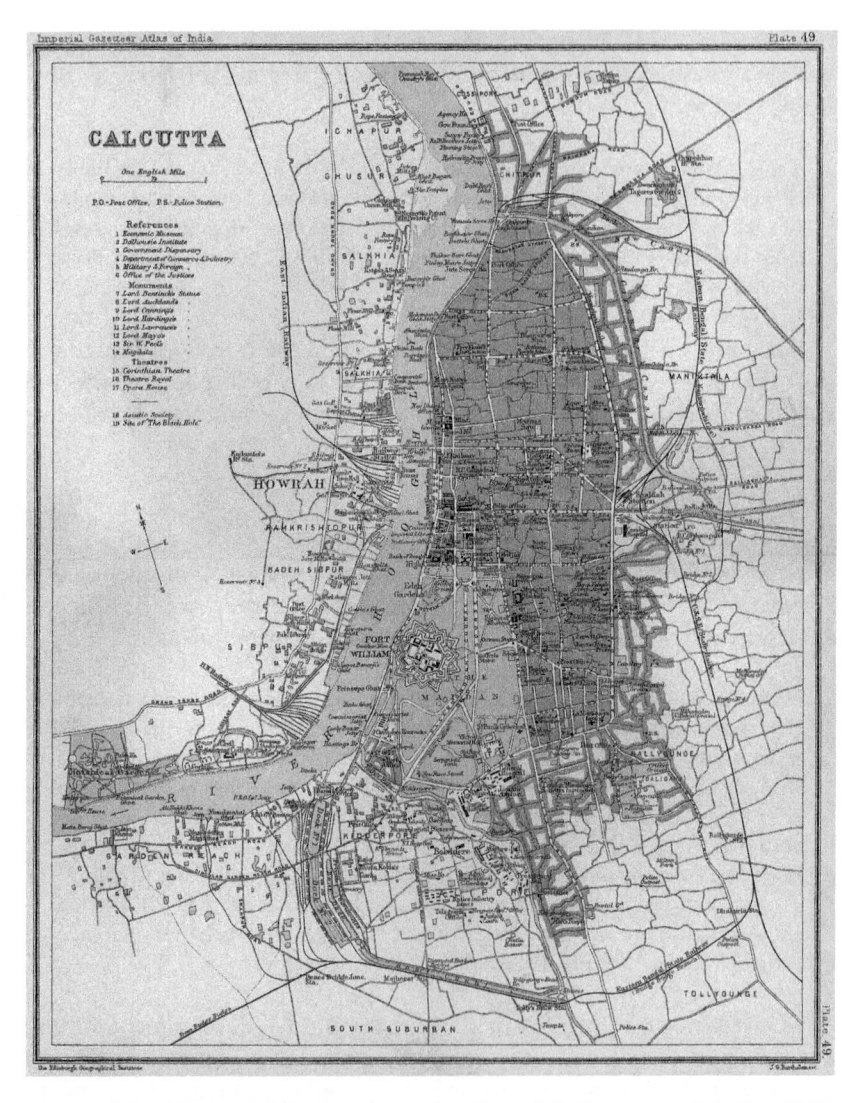

Image 2.1 Map of Calcutta from the *Imperial Gazetteer of India 1909*. Kidderpore is in the south where the river takes a westward turn. The warehouses and the *ghats* (landing places) dot the banks of the river on both sides, Calcutta on the east and Howrah on the west. (Image: Courtesy of the Digital South Asia Library, http://dsal.uchicago.edu)

the demands of the trade' (895). Many new additions to the dock complex were proposed, but the First World War halted the process because 'The war brought about a large reduction in the tonnage of vessels visiting the port, a huge drop in imports and the practical disappearance of the coal trade' (896). He further lamented that:

> A considerable portion of the plant of the port was commandeered for service in other ports of the Empire. The third suction dredger then under construction, a number of cranes, railway wagons, launches, and building material were all commandeered, and although the Trust received compensation, the net loss then incurred reached a very heavy sum, owing to the unprecedented and unexpected costs of replacement. (Stuart-Williams 1928, 896)

The post-war era put a lot of pressure on the Port Commissioners to replace the material taken during the war, but the cost of finance was almost 50–75 per cent higher than in the pre-war period. Up till 1921, few additions had been made to the Calcutta port and it took some time to restore it to its previous condition. As for trading activities, Nilmani Mukherjee in his history of the Calcutta port writes that in the post-war era, '[i]mprovement in trade conditions was painfully slow but the growth of trade was undisputed. In 1924–25 it was officially noted that the port was slowly regaining the old pre-war figures of general import traffic while it exceeded these figures in the case of general export' (Mukherjee 1968, 125). Also, a new phase of activity—including construction of four general berths at the Garden Reach area and of the King George's Dock—increased the accommodation facilities enormously. The new dock was opened in February 1929 and, despite a surge in trade during 1929–30, the Calcutta port was soon hit by the Depression, so that a considerable portion of the available accommodation remained unused during those years. Yet, although construction of additional berths was stalled, the tide seemed to turn during the mid-1930s when trade conditions improved and the Calcutta port benefitted from '[h]eavy imports of rice from Burma in 1934 and after, of steel and machinery and of Java sugar and Australian wheat in 1938–39 and the improvement in the shipment of coal, pig iron and manganese ore' (150). However, the Port Commissioners became concerned over the export of Indian sugar through the Calcutta port to other ports of

India because they knew this traffic had to compete with Java sugar. For this reason, they requested that the government keep port charges low for this sugar and the government obliged (NAI 1935, Marine Dept, Nos. 241–244). With the port authorities trying various methods to stabilize the trade situation, and it was important to augment the export trade, these steps took a further severe jolt with the outbreak of the Second World War. This time, the port of Calcutta was directly involved in the war effort of the Allied group, particularly after South-East Asia became a major theatre of the war. Then, considerable quantities of Army Stores occupied the berths at the docks and general trading activities were halted. Japanese air raids also proved detrimental from the point of view of labour, as some workers were killed and many deserted the dock area (Mukherjee 2014). This meant that the main problem for the Commissioners during the war-years was the slow clearance of goods that got accumulated in the berths and warehouses.

This ebb and flow of trading activities, connected with the general political and economic condition of India as a colony of the British Empire, shaped the ways in which the port complex developed in Calcutta. The infrastructural development of the port in this period gives us an idea of what Ned Rossiter (2016) has termed the 'proto form of logistical media' in the context of nineteenth-century development of telegraph and railways. These infrastructural programmes, which were initiated by the colonial states, Rossiter argues, 'produced the territorial imaginary of empire and economic system of imperialism' (150).

One of the crucial elements in any infrastructural growth project is the management of time and an efficient system is one where the amount of time spent in achieving a desired result is reduced. In the case of managing and functioning of the warehouses, as well as the building of new communication channels, the port authorities had to take into account the timely circulation of goods. Any study of the infrastructure thus, apart from interrogating the spatial aspect, also demands consideration of temporal dynamics. Therefore, in relation to the accommodation and transport facilities connected to the port, this chapter delineates the different mechanisms that were put in place to produce the space of the port. It includes micro-stories that provide a glimpse of the situation on the ground and give a sense of how grand strategies and big development projects were played out on the site.

Warehouses

By the mid-nineteenth century, a massive increase in trade indicated that more storage space was required in the port area of Calcutta. Between the 1850s and 1880s, the trade in jute, cotton, and tea increased rapidly and Calcutta became the main entry-point for imports of cotton piece goods. From the port, these were distributed throughout the provinces of Assam, Bengal, and parts of northern and central India (Bandyopadhyay 1995, 20). Due to the increase in tea exports during the 1870s, the Port Commissioners decided to build a tea warehouse on Strand Bank. Initially, however, the Bengal Chamber of Commerce criticized the step as it believed that the taking up of land on the Strand would interfere with private enterprises. Even so, while debate regarding the location of the warehouse ensued, the Commissioners went ahead with other issues associated with the building of the warehouse. In 1876, they asked all the mercantile firms who were involved in the tea trade to get back to them about designs for a suitable building and 22 firms responded positively to the entire scheme. A subcommittee, which was formed to oversee construction of the warehouse, met three times to discuss the building plans, the mode of working, and the scale of charges. A circular was issued with the proposed scale of charges, and the Commissioners asked the firms whether they would still be interested in the trade if the charges were levied in the warehouse. While the tea-brokers remained adverse to the entire scheme from the outset, the firms were more or less in favour of the project going ahead (Administration Report of Port Commissioners 1877, 31). After various contestations and negotiations, the tea warehouse at the Armenian Ghat, situated at the north of the Fort William along the Strand Road, was ultimately made available from 1887.

The Port Trust from its inception had to deal with the issue of private property and its acquisition. The facilitation of maritime activities involved taking up extensive swathes of land along the river banks and, with this, the port story slowly moves into the larger narrative of urban governance. Town and port authorities combined to formulate rules and regulations, and a brief survey of the legislative history regarding warehouses and allied issues offers insight into the ways in which the provision of storage space was conceived and put into practice.

Under Act XXV of 1836, the Governor of a Presidency could declare any port within his territory as a 'warehousing port.' The act made a distinction between 'public' and 'private' warehouses and stated that 'the

Warehouse of the Custom House, together with such other Buildings as shall be directed by the Governor in Council, or Governor of the Presidency, or Settlement, shall be Public Warehouses for the reception of the Goods under the provisions of this Act.' It further ruled that 'every Public Warehouse shall be under the lock and key of the person whom the Governor, or Governor in Council of the Presidency, shall appoint to be responsible for all duties connected with the charge of Goods, their reception into, and delivery from the Warehouse' (Bengal and Agra Annual Guide 1841, 161). The Act issued orders to private warehouses that they had to obtain licences from the government to be able to operate in the business of the ports and, to have their licence granted, they had to follow a series of procedures and regulations (158–64). These addressed crucial issues that would determine the fate of large chunks of privately owned properties in the city and one of the most important of these was fire safety. The Licensed Warehouse and Fire Brigade Act 1893 provided for the levy of a special taxation for the maintenance of the Calcutta Fire Brigade. The taxation was in the form of licences for warehouses used for the storage of inflammable goods. The Corporation of Calcutta issued the licences for the warehouses, which was done after consulting the Commissioner of Police. The amount was determined as an annual fee not exceeding 10 per cent of the annual assessment (Report of Bengal Chamber of Commerce 1913, 55). The Act, however, divested the Calcutta Port Commissioners of all responsibility for the control and administration of the fire brigade. They were also not required to inspect and supervise the warehouses. With this, the jute department which was till then maintained as a branch of the licence department was abolished (Goode 2005 [1916], 282). Previously, when large premises in Darmahatta and Armenian Streets were burnt down in 1871, the Justices of Peace were forced to look into the condition of warehouses storing inflammable items like jute and cotton, and to maintain the fire brigade on a more efficient footing. Under the Jute Warehouse Act II 1872, the licensing of warehouses was made more stringent: regular magisterial inspection was ordered and the various municipalities of the town and suburbs were ordered to maintain an efficient fire brigade, with the cost of the fire brigade charged to the individual municipalities. Apart from the fees levied on the jute merchants, 'a rate was realized from the Fire Insurance Companies' which, as the municipal historian S. W. Goode mentions, was 'calculated upon the amount of premia received by them. The amount raised by these means was large enough to enable Government between 1872 and 1881 to expend more than 1¼ lakhs of rupees out of the surplus of the Fire Brigade Fund on works of public improvement' (282–83). The Act V of 1879 moderated the severity of the assessments by including more items as taxable material. Under this Act, the Jute Warehouse Fund was established and the town and suburban commissioners could use this fund to maintain the fire brigade and pay all expenses for the inspection and supervision of the jute warehouses. In response, in 1890, the Bengal Chamber of Commerce protested against this Act, as it was seen that the burden fell almost entirely on one industry—jute. A committee was appointed to look into the problem and, finally, the Licensed Warehouse and Fire Brigade Act 1893 was passed where only half of the annual cost of maintaining the fire brigade was to be derived from the licence fees of the warehouses, the other half would come from municipal revenues (283).

Issues of safety and security measures undertaken in the warehouses were crucial for business. Often fire would destroy a large quantity of goods (*Times of India*, 20 December 1907; 21 October. 1931). Also, erosion in the river bank created cracks in the stone foundations as well as the walls of the jute and tea warehouses (*Times of India*, 2 March 1955). The successive acts regarding safety from fire repeatedly take note of the condition of the warehouses in the city. Through a series of regulations, the licence system and taxation, the town authorities tried to maintain the functioning of the warehouses and, as trading activities increased, port and town authorities had to make provisions for the safety and security of the warehouses, as well as keeping an eye on revenues.

With the formation of the Port Trust, many new warehouses were soon proposed. A jute warehouse was planned on a portion of Strand Bank land between Ahiritolla Ghat and the Mint in 1872. The advantages of this particular site were eloquently articulated by the Commissioners of the Trust to the justices of the town and the government. The port authorities wrote:

> The lands are separated from the town by the Strand Road, and are thus so isolated as to ensure comparative safety to the town buildings in the event of fire originating in the proposed warehouses. Having a river frontage, and on the land side the tramway, when [it] is to be constructed in connection with the whole municipal system and with the Eastern Bengal Railway, there is every facility for the easy conveyance to the site of all raw material brought to Calcutta either by the Eastern Bengal Railway or by river steamers and flats, and for removal of exports when prepared and ready for shipment by the tramway, which will be in direct communication with the jetties. (NAI 1873, Financial Dept, Nos 55–57)

The reasons were compelling enough and the plan was sanctioned by the government. A loan of Rs. 2 lakhs (Rs. 200,000) was granted to the Commissioners to construct the warehouse. We find here a description of a complex system of communication with water, roadways, tramlines, and the railways. While the road would cordon off the site from causing damage to the town in case of an accident, the place was also situated favourably to connect with 'the whole municipal system' and beyond. The Strand occupied a crucial location in the city, connecting as well as separating the river and the city, the worldwide business of empire, and the everyday rituals of the pious population in the holy river.

The interests of various groups came into conflict with each other in developing the port complex of Calcutta and the government had to regularly take into account the considerations of the business community before embarking on a port-related project. The government in Bengal had been contemplating the construction of wet-docks from the 1830s. Diamond Harbour, almost 50 kilometres south of Calcutta, was thought to be an ideal spot. In the 1880s, the scheme was given fresh impetus, but the Diamond Harbour wet-dock scheme was opposed by Calcutta's mercantile community. They argued that it would require additional investment on their part to transfer the current activities to another port and they wanted the government and the Port Trust to first look into all the suitable spots, in and around the Calcutta port, for additional space on which to build the wet-docks. Finally, their opposition proved vital in initiating the construction of the wet-docks at Kidderpore (Report on the Construction of Docks 1885, 199). But, as the Commissioners' report suggests, more pressing than the wet-docks was the need for additional storage space in the vicinity of the Calcutta port. A committee that was formed to look into the provision of a railway junction and a bridge over the Hooghly stated:

> As regards the convenience of the trade of Calcutta, there seems little to choose between any site along the canals and Circular Road, from Chitpore to Sealdah. The main business in warehousing ... is carried on in the part of the city bordering onto the Hooghly between the Custom House and Chitpore, and all points on the line named would be nearly equidistant from the centre of this class of business.
>
> It is a fact sufficiently attested that the trade of Calcutta, as now conducted, requires that the mass of the goods for export (which form the most important part of the goods dealt with by the Railway Companies) shall be

re-packed in Calcutta. This involves their delivery by cart at the warehouses of dealers. Probably the formation of wet docks, with warehouses attached, might hereafter, in some measure, change the habits of the trade, but meanwhile the requirements of the existing state of things must be met. Hence a large ordinary goods station must be formed, suitable for the present condition of business, quite irrespective of the question of docks. (58)

The committee appointed by the Commissioners in 1881 suggested that a new line of warehouses were needed to be constructed as soon as possible to ease the heavy pressure of increased goods in the port. Limited covered space was available in the port area, and that caused many problems for business. The report mentioned that 'the sheds become crowded with goods almost immediately after a ship commences her discharge; and when two ships occupy the berth one after the other, each bringing a large cargo, the work of sorting and delivery becomes most difficult, and is the cause of frequent complaints' (Administration Report of Port Commissioners for 1881–82, 1882, 22). New warehouses would, as the report noted, 'enable the Calcutta jetties to meet the demands of a growing trade, and compete on more equal terms with the appliances existing in other ports' (22). In February 1882, the first block of warehouses at No. 1 jetty was commenced. Another block was sanctioned by the government in 1882 at No. 3 jetty, which was entrusted to Messrs S. C. Mitter and Company (Administration Report of Port Commissioners for 1882–83, 1883, 3). But before giving the go-ahead, the government had its doubts. The British Indian Association and the municipality feared that the new warehouses would diminish the value of privately owned resting sheds in the city. The proposed project of the Commissioners seemed to give the impression that the new warehouses were being built for the purpose of renting them to the merchants and traders, opined the Lt-Governor. The Commissioners hastened to dispel any misconception and replied 'that the new warehouses were intended to supplement and relieve the existing jetty sheds, and that there was no intention to rent them out for business unconnected with the landing or shipment of goods through the jetties' (22). In the opinion of the Commissioners, regular importers or exporters would find it convenient 'to rent a certain space in the new warehouses for the storage of their goods pending dispatch or shipment instead of keeping them in the ordinary jetty sheds where examination and assortment of the goods was rendered difficult in consequence of the goods of different firms being mixed together' (22). The Commissioners contended that this use of warehouses 'was a legitimate one and was in accordance with the practice in all large ports' (22).

The issue of private resting sheds was not the only obstacle. The Municipal Commissioners of the town raised an objection regarding building an elevated structure on the Strand Bank, following the instruction given by the Governor of Bengal in 1852 when this piece of land was acquired by the government for public utility. The Lt-Governor, however, thought that the warehouse was a necessary structure for the advancement of trade and was not antithetical to the use of that piece of land for the good of the general public. These objections and negotiations reveal the difficulties associated with the initial phase of the construction of warehouses in the port of Calcutta. Issues regarding private property, proper use of land, trade charges, backing from the mercantile firms, and the views of the Port Commissioners about modern port facilities jostled with each other in regard to the establishment of warehouses.

However, we must also keep in mind that often it was not just accommodation at the port that was an issue; rather, the dues charged on merchandise also became a crucial factor for trading activities. In 1885, various merchants, mill owners, and jute balers wrote a letter to the Lt-Governor of Bengal regarding the bill in the Bengal Council that gave the charge to the Port Commissioners to build docks at Kidderpore and to raise loans for that purpose. They thought that at present it was not necessary and they pointed out that 'the export trade of Calcutta has lately shown unmistakable signs of falling off, and that what is required at present is not so much additional docks and jetties, as that the charges of the Port should be decreased to enable Calcutta to hold its own against Bombay and Sindh' (NAI 1885b, Public Works Dept, Nos 1–4). The merchants noted that considerable additions were made at the Howrah terminus of the East Indian Railway which enhanced the prospects of export trade enormously and that increased accommodation in the port would only be needed if export trade grew. But with the proposed dues to be levied in the new dock, the merchants believed that, instead of facilitating the growth of trade, the dues would in effect render useless any additional space as the cost of export would increase enormously. The chief items of export at that point of time were jute, wheat, rice, gunny bags, and oilseeds. These might need further space in the docks but, if trade decreased due to the increased rate of customs, any new development at the port would be practically of no use. For instance, in the case of jute, the merchants mentioned that 'the present practice [was] for the raw fibre to be pressed into bales, at different press houses on the river bank, and for it then to be loaded into cargo boats, and sent alongside the export ship, at a cost of from 10 annas to Re. 1 per ton' (NAI 1885b, Public Works Dept, Nos 1–4). They feared that this system would be under threat when the new dock with increased dues started functioning and argued that 'As the dock dues proposed to be levied on this article are Re. 1 per ton, it is obvious that so far from being a boon, the docks will increase the charges on this fibre.' Trade in gunnies and rice would also face similar problem. For these merchants and jute balers, a reduction in the charges was more desirable than any increase in facilities. In their opinion, the trade handled at the port did not warrant any extension at that point of time, rather 'the building of a dock at an enormous expense will be a great burden on the trade of Calcutta, because it has never been shown that a dock will be the means of either reducing charges or facilitating dispatch.'

Notwithstanding such objections from a section of the trading community, the port authorities always looked to acquire more space for various activities. In 1881, the Port Commissioners proposed to purchase the property belonging to the Calcutta Docking Company, which was situated on the Howrah foreshore, north of the Hooghly Bridge. The Government of India was also keen on the project as it needed a space to store materials for the railways. The project did not materialize as the company directors, on behalf of the shareholders, did not accept the amount of Rs. 450,000 offered by the Commissioners. The government also did not pursue the matter. But the Commissioners decided to take the matter up again due to the great inconvenience in carrying out the docking and repairs of several vessels belonging to the Port Trust. Apart from ships and vessels, there were large quantities of materials belonging to several departments that laid scattered in various locations of the port. The Commissioners asked the Calcutta Docking Company the price of the land and they were told they wanted Rs. 575,000. The Port Trust made an offer of Rs. 500,000 and the parties ultimately settled for Rs. 525,000 (NAI 1882, Dept of Commerce and Industry, Nos 1516–18).

In 1912, the government approved the building of a two-storey warehouse on the foreshore of river Hooghly, north of the Howrah Bridge, for the convenience of the inland vessels companies. A revision of the earlier plan was done and soon it was found that, with minor alteration in the alignment of the proposed warehouse, it could be built on a larger area. While the changed location would facilitate better connectivity with the railway lines, it would also mean an increase in the budget. By this time, however, the port of Calcutta had gained immense importance in the imperial trade network and so the government did not hesitate to sanction the extra amount required for building the larger warehouse on the other side of the bank in Howrah (NAI 1912, Dept of Commerce and Industry, Nos 8–9).

Thereafter, trading activities at the port continued to increase, with some occasional setbacks during war or the Depression years. More jetties and warehouses were ordered, and the Port Commissioners continuously put pressure on the government to maintain adequate funds. In 1895, the average daily imported goods weighed around 1000 tons of which 300 tons were stored at the warehouses, with the rest being carted away to other parts. In times of high demand, this amount doubled (*Times of India*, 4 June 1895). In fact, a decade later, in 1906, the Secretary of the Bengal Chamber of Commerce mentioned that due to the increase of trade, there was hardly any space at the jetties for the imported goods. He noted that new jetties and a modern crane system were being constructed at the port, but these were not enough and new warehouses were needed. While at that time, a new tea warehouse was being constructed in Garden Reach which would help in opening up almost 15,000 square feet of space at the jetties for import trade. The Secretary wanted the port authorities to construct a new warehouse for import trade on Strand Road frontage (*Times of India*, 8 February 1906). This complaint of shortage of space in tea warehouses was a recurrent feature in the first half of the twentieth century (*Times of India*, 25 October 1939). After independence, the new government also faced this problem, and an ad hoc Committee was established in 1950 to look into the matter, although, at the time, it was reported that the Port Commissioners of Calcutta were constructing a four-storey permanent tea warehouse with floor-space of 140,000 square feet and a tea transit shed covering 20,000 square feet between the present tea transit sheds nos 1 and 2 and the sales tea warehouse (*Times of India*, 8 April 1950). Thus, over more than half a century, the demand for storage space to meet the ebb and flow of trade volume shaped the port complex in Calcutta, and the exigencies of trade, global warfare, domestic demand, and pressure from mercantile firms all combined to mould port activities and infrastructure.

An important aspect of the warehouses was the operations that took place inside them. A major concern, for example, was the proper measurement of weight of the goods and disputes often arose regarding the method of weighing. For instance, in 1901, the Indian Tea Association sent a letter to the Port Commissioners urging them to broach the fact to the government that the English Customs Department should accept the weight of the tea ascertained by the Calcutta port authorities in their warehouses. The Port Commission agreed to this proposal and urged the government to look into the matter. They gave a detailed description of the process of weighing, arguing that there was very little chance of any error and no loss could possibly accrue to the English revenues if they accepted the weight as measured in Calcutta. The process described was as follows:

> The tea having being bulked in the patent machine which the Commissioners have erected, passes by gravity into the weigh hopper. From this hopper the required contents of each chest is weighed and discharged by gravity into the chest, the tare of which has been ascertained by separate weightment. The loose tea is then compressed into the chest by hydraulic power and the chest is closed and the gross weight taken, which is checked by the already ascertained tare and the weight of the tea put on to the chest. (NAI 1901, Finance and Commerce Dept, Nos 240–41)

The concern over weight and measures was persistent. In 1950, the ad hoc Committee formed to look into the problems of the tea trade noted that, in the tea warehouse, only 10 per cent of the product, randomly chosen, was inspected. This was deemed inadequate for ensuring the quality of the tea or the security of the packaging and 100 per cent inspection was recommended. For this, additional warehouse space was needed as that would help with the packaging and handling of tea chests as well as inspections (*Times of India*, 8 April 1950) (Image 2.2).

TRANSPORT

The Port Trust had initiated large-scale infrastructural development during the 1870s. One of the major areas of interest was to create a proper channel of transport facilities to move goods to and from the dock area. The railways played a crucial role in connecting Calcutta with other parts of the province and country, and Calcutta was served by the East India Railway, the Bengal-Nagpur Railways, and the Eastern Bengal Railways. The development of the railways was crucial in facilitating the port's

Image 2.2 Balance scale in a warehouse near Howrah, 1944. (Photograph by Glenn S. Hensley, held by University of Chicago Library, Southern Asia Department)

activities and the major terminals were at Sealdah (Eastern Bengal Railway) on the east of the river on Calcutta side, and the other was on the west at Howrah. In the 1880s, a bridge was proposed to be constructed over the Hooghly. The Lt-Governor of Bengal in 1883 noted that:

> The future developments of trade which the continual progress of railways encourages are incalculable; and when the bridge over the Hooghly is finished, and direct communications with Calcutta have been established from the producing districts of the North-Western Provinces, and the tracts of country served by the Northern Bengal, the Central Bengal, and the contemplated line from Seetarampore to the Central Provinces, with their connected branches, the space at present at the disposal of the Port Commissioners seems to me to be utterly inadequate. (Report on the Construction of Docks 1885, 158)

Networks of railroads promised brisk business for the Calcutta port. While major items like rice, coal, and jute were transported to other parts of the subcontinent from the port via the railways, in the immediate vicinity of the port, proper roads and carriers were unsuitable for the handling of a large bulk of cargo. To address this, the Port Trust began constructing a tramway along the Strand. The tramway work progressed rapidly with materials being imported from England. In their report of 1877, for instance, the Port Commissioners mentioned that the trust had been able to obtain a burning *ghat* (crematorium) site and section no. 17 of the new road between Ahiritolla and Ruth Ghats which enabled them to complete the work as far as the Armenian Ghat in the north. The trains ran daily bringing the cargo from the Eastern Railway to the freight sheds on the inland vessels wharves. The development of the tramways was directly linked to the massive increase in net cargo handling in the port and the successive stages of tramway construction gives us insight into the gradual extension of port activities and the way crucial links were established between the docks and the city, and in turn with the hinterland. Various new plans were proposed and some were followed while negotiations on ground forced a few changes and alterations. For instance, the Commissioners noted in 1877 that 'the traffic passes over the municipal line of railway from Sealdah to Bagh Bazar; but this is only a temporary arrangement, the Commissioners having ... undertaken to construct a bridge across the entrance to the Chitpore Canal, and so carry their line of tramway direct into the Eastern Bengal Railway goods terminus at Chitpore' (Administration Report of Port Commissioners for 1876–77, 1877, 3). To use the municipal line the port authorities had to enter into an agreement with the Town Commissioners and the terms of the agreement included the following:

1. That the Port Commissioners shall pay eight annas per wagon for every wagon that passes over the municipal line, either way, full or empty;
2. That the Port Commissioners shall have free use of the line for six hours daily, from 7 to 10 am in the morning and 3 to 6 pm in the afternoon;
3. That the Port Commissioners shall pay the cost of keeping that portion of the municipal line over which the trains run in repair;
4. That either the Town or Port Commissioners shall have the option of terminating the arrangement by giving one month's notice at the end of each year after the second year.
5. That this arrangement shall be binding on both parties for two years certain. (3)

With this arrangement with the railways and the town authorities, the port tramway was inaugurated on 22 November 1876. But crucial works remained to be done. Originally, the intention of the trust was to carry the tramway line across the mouth of the Chitpore canal by building a moveable bridge. However, objections were raised as it was feared that it would interfere with the traffic on the canal, and the 'Government required that any bridge to be constructed in this position should have a clear headway of 16 feet above high water. To obtain the necessary incline for the approaches to such a bridge, an embankment would have to be made at the frontage of the Eastern Bengal Railway Company, which would shut out the Company from access to the river, and to this the Company would not have agreed' (3). Also, an elevated line would cost around Rs. 4.5 lakhs (Rs. 450,000) which was not possible to recover from the goods traffic on that line. The Commissioners decided to abandon that route as they thought that a fixed bridge was the only solution and a new bridge was designed, keeping in mind all the objections of the canal authorities while providing a passage for the trains at ordinary level, at a cost of about Rs. 90,000.

Apart from the bridge, a major problem also arose with the connection of the jetties with the inland wharves and the Eastern Bengal railway line. The proposed tramway was passing through the Armenian Ghat station, and the East Indian Railway Company did not agree to dismantle the station. A long-drawn negotiation ensued. The Strand tramway line was of immense importance to the port as well as for other departments, especially for the army headquarters at the Fort William. The line ran along the boundary of the fort on one side and thus provided an excellent opportunity for military stores to be carried by government wagons. The port authorities did not have any objection to such usage of the new line as long as the control of the traffic on the line was to rest with only one authority—the Port Trust. The Vice-Chairman of the Commissioners for making improvements in the port of Calcutta informed the Brigadier General in command of the Presidency District:

> There will be no objection to Government using its own wagons for the conveyance of stores, and moving such wagons either by steam or manual labour on the sidings leading into the Government premises, but there are serious objections against the main line being used by any other engines than those belonging to the Commissioners. It would be impossible to regulate the traffic over the main line if Government and the Commissioners both had the power to move wagons along it whenever they pleased, and would certainly lead to some serious accident. The haulage of wagons must … be done entirely by the Port Commissioners' engines in the same manner as the traffic is worked at present between Cossipore and the Jetties, where all wagons whether belonging to the Port Commissioners or to the Government State Railways are hauled by the Port Commissioners' engine. There is no objection to Government wagons being used to any extent, but the engines on the line must be under the authority of the Commissioners, or they could not be responsible for the safe working of the line. (NAI 1885a, Public Works Dept, Civil Works-Misc. Branch, Nos 1–3)

The tram lines soon became profitable. Between 1880–81 and 1882–83, there was an increase of almost Rs. 15,000 in tramway receipts (Banerjee, 1975, 41). Also, the increase in traffic necessitated opening up a third line (with already two lines for up and down traffic inaugurated in 1881) between Nimtollah Ghat and Ruthghat within a year of its functioning (40). But, we must remember, the final haul, from the wagons to the warehouse, was done by human labour and, as the following photographs (taken in 1944) show, these "human conveyor belts" were essential in placing the goods inside the warehouses (Images 2.3 and 2.4).

Images 2.3 and 2.4 'Human Conveyor Belt'. (Photograph by Glenn S. Hensley, held by University of Chicago Library, Southern Asia Department)

The development of the roads and tramways gives us a glimpse of the manner in which the port area was extended and integrated with the rest of the city, the difficulties that arose regarding land or finance, the negotiations that ensued between various branches of the government, and the general implications of ongoing expansion for the trading activities of the Calcutta port.

Coda

Let us end with the story of Burma rice during the First World War. Rice produced in Burma was exported to India through Calcutta port and railway wagons carried it to other parts of the country. This rice was usually cheaper than Indian rice and was consumed by a section of the poor. In 1917, during the war, the rice situation in Burma was facing a crisis. There was abundant production, but a few dealers held onto it in the hope of a rise in the price. In the meantime, wagons to supply coal from Calcutta were needed by the railways and it was proposed that, as there was no need to import Burma rice, the wagons should be used instead to transfer coal as the smooth operation of the railways was deemed to be 'a matter of Imperial importance' (NAI 1917, Dept of Revenue and Agriculture, No. 7). There had been good crops in India that season and so, apart from the difficulties with freights, the Indian authorities were not keen to accept more rice from Burma in light of its own abundant produce. This decision put a lot of stress on the rice industry in Burma. While the port authorities in Calcutta also said that they did not have enough storage space to keep the rice under their control, the Burmese government tried to point out that there was a demand for Burma rice in some of the upcountry provinces in India. They were ready to divert their export trade through any port other than Calcutta owing to the congestion and objection regarding storage facilities and transportation at the port. However, they were clear 'that such steps should be taken only in the last resort' (NAI 1917, Dept of Revenue and Agriculture, No. 7). In the end, to ease the situation in Burma, the home government in Britain agreed to import some quantity of rice.

In fact, congestion in the port area was raised as a recurrent feature which meant that, on occasion, some goods had to be prioritized and the problem with Burma rice arose again in 1919. The authorities knew that it was essential that sheds in the dock be kept clear for the arrival of the rice, and that the rice shipment should be regulated in keeping with the railway schedule. If not, there would be undue congestion in the docks and the railways would not be able to carry off the consignment. In the case of sugar from Java and wheat from Australia, the preferred destination was always Calcutta port and, although the port authorities lacked the space to store this produce, they did not want to lose out on the customs revenue. The Port Commissioners requested the railway board to provide them with more warehouse space, as sometimes goods were stored by the port authorities in the sheds at Howrah belonging to the East Indian Railway, but on this occasion, the railway board refused to grant any more space to the port traffic. Ultimately, this meant that special arrangements had to be made to supply a large number of wagons to the port authority to clear the stock of sugar and rice so that fresh imports could arrive (NAI 1919, Dept of Commerce and Industry, No. 25, 1919).

The activity in the docks needed to be regulated and systematized so that a smooth functioning was possible. It was not only a question of storage or increased trade, equally important was the management of time, scheduling the movements, and the dispersal of the goods. The story of the Burma rice also reveals the way in which the Calcutta port was intricately linked with the imperial traffic of commerce and war. Timothy Mitchell (2014) has suggested that, 'infrastructures are both durable yet fragile, hidden but ever present, solidly embedded in the collective world yet open to speculation and uncertainty' (437). The fragility or uncertainty of infrastructures reveals their underlying architecture. The history of building, functioning, and maintenance of this infrastructure opens up ways of looking at how the life of this infrastructure sustains wider networks of trade, energy, or public health services.

This chapter has traced the development of port facilities in Calcutta with the increase in trade from the 1860s to the 1950s. It has studied two aspects of this—storage and transport—both of which are related to the massive increase in the bulk of cargo. Related to these individual systems (of warehousing and transport) was the production of the space of Calcutta port where different actors—the Port Commissioners, municipal authorities, imperial government, mercantile firms, or the railways—all staked a claim. Through an analysis of the infrastructural development of the warehouse and the port tramway, and the negotiations and contestations they entailed, I have tried to capture the evolution of their internal workings, management, and safety measures. These activities were intimately connected with the wider political and economic scenarios including global wars and depression. The port of Calcutta was a crucial node in the British imperial network and this chapter has also explored the bearing this had on the development of the port complex of Calcutta.

REFERENCES

Administration Report of the Commissioners for Making Improvements in the Port of Calcutta, 1876–77 to 1882–83. Calcutta: Bengal Secretariat Press.
Ahuja, Ravi. 2009. *Pathways of Empire: Circulation, 'Public Works' and Social Space in Colonial Orissa (c. 1789–1914).* Hyderabad: Orient Blackswan.
Banerjee, P. 1975. *Calcutta and its Hinterland.* Calcutta: Progressive Publishers.
Bandyopadhyay, A. 1995. 'Realms of Imperialism.' In *Port of Calcutta: 125 Years, 1870–1995, Commemorative Volume,* edited by S. C. Chakraborty. Calcutta: Calcutta Port Trust.
Bengal Chamber of Commerce. 1913. *Report of the Committee of the Bengal Chamber of Commerce for the year 1912.* Calcutta: Criterion Printing Works.
Goode, S. W. 2005 [1916]. *Municipal Calcutta: Its Institutions in their Origin and Growth.* Calcutta: Macmillan.
Mitchell, T. 2014. 'Introduction: Life of Infrastructure.' *Comparative Studies of South Asia, Africa and the Middle East* 34, no. 3: 437–9.
Mukherjee, J. 2014. 'Japan Attacks.' In *Calcutta: The Stormy Decades,* edited by T. Sarkar and S. Bandyopadhyay. New Delhi: Social Science Press, 93–120.
Mukherjee, Nilmani. 1968. *The Port of Calcutta: A Short History.* Calcutta: The Commissioners for the Port of Calcutta.
National Archives of India, New Delhi. 1935. Marine Department, Proceedings B, Nos 241–44, January 1935.
———. 1873. Financial Department Proceedings, Accounts 'A,' February 1873, Nos 55–57.
———. 1882. Department of Commerce and Industry, October 1882, Proceedings Nos 1516–18.
———. 1885a. Public Works Department, Civil Works-Miscellaneous Branch, Proceedings A, January 1885, Nos 1–3.
———. 1885b. Public Works Department, Miscellaneous, Proceedings, August 1885, Nos 1–4, Part B.
———. 1901. Finance and Commerce Department, Statistics and Commerce Branch, Proceedings, May 1901, Nos 240–41.
———. 1912. Department of Commerce and Industry, Proceedings B, December 1912, Nos 8–9.
———. 1917. Department of Revenue and Agriculture, Proceedings, April 1917, No. 76, Part B.
———. 1919. Department of Commerce and Industry, March 1919, No. 25.
Public Works Department. 1885. *Report Connected with the Project for the Construction of Docks at Calcutta.* Calcutta: Superintendent of Government Printing.
Rossiter, Ned. 2016. *Software, Infrastructure, Labor. A Media Theory of Logistical Nightmares.* New York and London: Routledge.

Stuart-Williams, S. C. 1928. 'The Port of Calcutta and its Post-War Development.' *Journal of the Royal Society of Arts* 76, no. 3948: 890–906.
The Bengal and Agra Annual Guide and Gazetteer for 1841, Vol. 1, Part II. Calcutta: William Ruston and Co.
The Telegraph. 2010. 'Strand, a Picture of Tragic Grandeur.' 29 May. Accessed 14 October 2017. https://www.telegraphindia.com/1100530/jsp/bengal/story_12503066.jsp
The Times of India. 1895. 'Shipping Accommodation in Calcutta.' 4 June,
———. 1906. 'Calcutta Import Trade: Increasing the Accommodation.' 8 February, 6.
———. 1907. 'Bio Jute Fire: Warehouses Destroyed.' 20 December, 8.
———. 1931. 'Jute Warehouse Ablaze: Several Lakhs Damage at Cossipore.' 21 October, 7.
———. 1939. 'Tea Warehouses of Calcutta: Question of Congestion.' 25 October, 9.
———. 1950. 'More Warehouse Space for Tea: 'Ad Hoc' Committee's Recommendation.' 8 April, 4.
———. 1955. 'Erosion Affects Warehouse: Contents Cleared.' 2 March, 7.

CHAPTER 3

Spatialization of Calculability, Financialization of Space: A Study of the Kolkata Port

Iman Mitra

Located on the left bank of the river Hooghly, at latitude 22°32′53″N and longitude 88°18′5″E, the Kolkata Dock System (KDS) is one of the oldest dock systems in India. Its vast hinterland includes West Bengal, Bihar, Jharkhand, Uttar Pradesh, Uttarakhand, Madhya Pradesh, Chhattisgarh, Punjab, Haryana, Rajasthan, Assam, the north eastern states of India, and two landlocked neighbouring countries, namely Nepal and Bhutan. Currently, it has two approaches from the Bay of Bengal: the eastern and the western channels. Navigation to and from the port, at this moment, is only being done through the eastern channel, which is one of the longest navigational channels in the world. The pilotage distance to Kolkata is 223 km, of which 148 km is river pilotage and 75 km is sea pilotage. There are several navigation aids provided by the Kolkata Port Trust (KoPT)— the port management authority in Kolkata—for the safe passage of vessels: two lighthouses on Sagar Island and Dariapur on the right bank of Hooghly; 5 unmanned light vessels on the sea; automatic tide gauges

---

I. Mitra (✉)
Centre for Development Practice and Research, TATA Institute
of Social Sciences, Patna, India

© The Author(s) 2018
B. Neilson et al. (eds.), *Logistical Asia*,
https://doi.org/10.1007/978-981-10-8333-4_3

maintained at Garden Reach, Diamond Harbour, and Haldia for round-the-clock recording of tidal data; manual tide gauges maintained at Akra, Moynapur, Hooghly Point, Balari, Gangra, and Sagar; 500 river marks, 90 lighted buoys, and 42 unit buoys; a wireless very high frequency (VHF) network for communication between approaching vessels and onshore and offshore KoPT establishments and vessels; the electronic position fixing system 'Syledis,' and a satellite-based Differential Global Positioning System (DGPS) (Kolkata Port Trust 2016a).

As one can see, even a short description of this site evokes an entangled framework of infrastructural accumulation and logistical governance. This chapter seeks to understand this framework from two specific yet interconnected perspectives, that is, the spatialization of calculability and the financialization of space. The Kolkata (erstwhile Calcutta) Port is a site where these two perspectives collide and communicate with each other and give birth to a particular form of logistical governance. Although it could be argued that such arrangements apply in almost all the ports in the world, I shall try to illustrate how the Kolkata Port interprets and dismantles some of the standard elements of logistical governance. This form of governance requires negotiations with, and navigations through, a network of institutional apparatuses which produce the material basis of calculations and speculations that envisage the connections between infrastructure and logistics. I will show that logistical governance in the Kolkata Port rests on the particularities of correspondence between institutional apparatuses like the KoPT (the semi-autonomous management authority which runs the port) and specific regimes of calculability and speculation.

Setting the Framework

Before getting into the details of the port as a site of logistical governance, let me explain what I mean by the terms 'spatialization of calculability' and 'financialization of space.' I am borrowing the term 'calculability' in part from Timothy Mitchell who, taking a clue from Georg Simmel's writings, points to the essential correlation between the conception of modern life as governed by endless calculations and the politics of knowledge production that has 'space' at the core of its realization (Mitchell 2002, 80–119). The spatial order that has come to be associated with this knowledge is 'relational,' but not just in the sense in which David Harvey uses the term when he describes 'relational space'—that is, space 'regarded...as being contained *in* objects in the sense that an object can be said to exist only insofar as it contains and represents within itself relationships to other objects' (Harvey 2009, 13; emphasis in the original). Here, the issue at stake is not so much one of recognizing the relational ontology of space but one of producing these relationships through a complex of calculational and representational techniques. These techniques, as Mitchell suggests, work as parts of the governmental machine to spot various 'irregularities'—the repairing of which, then, becomes the task of government.

This conception of 'calculability' is essentially linked with space since the relational investigation and repair of irregularities are possible only within strictly defined spatial coordinates. If we transport the concept to a neoliberal context, however, a different picture emerges whereby irregularity does not always imply an entirely disadvantageous situation as, more often than not, neoliberal capitalism operates by cultivating unevenness; for example, in terms of flexible labour laws, exceptional fiscal reliefs, extraordinary bailouts, and other forms of governmental assistance. In such cases, there is a seeming role-reversal, in which confidence in an interventionist paradigm of governance undermines the neoliberal faith in an auto-corrective market. Such interventions also require a demarcation of spatial coordinates, ranging from the old but still persisting division between the North and the South to the designation of 'special' economic and financial zones in the post-colonies. Capitalism in the twenty-first century is not only a story of accumulated wealth; increasingly, it is also a saga of informalization of the economic sphere.

Given these changes during the last two decades, we need to extend the concept of 'calculability' as well. Instead of defining it as a mechanism of finding and repairing irregularities, we may think of it in terms of the management of uncertainties. There are various forms of uncertainty, many of which result from the hegemony of speculative capital flows and the circulation of immaterial goods. In the context of the present study, however, I am more interested in another type of uncertainty which is related to the politics of space itself. Space-making as a material practice involves two types of activities under neoliberalism: zoning and debordering. On the one hand, we witness a dismantling of borders and relaxation of boundaries to ensure the free movement of resources; and on the other, we see numerous attempts at the concentration of these resources within zones that are deregulated and informalized (but not ungoverned). In effect it seems, both modes of space-making are subject to uncertainties that are unavoidable and, at times, indispensable. But, more importantly, the struggle for command over the spaces produced thereby is delimited by the fervour with which these uncertainties are governed.

Governing uncertainties has been one of the motivations of the liberal regime of calculability as well, but that too was driven by a wish to get rid of irregularities. Conversely, the neoliberal regime of calculability champions the existence of irregularities which are space-bound and instrumental to capital formation and accumulation. This axis of spatialization of calculability is incomplete without another phenomenon: financialization of space. The real-estate boom in the last two decades has turned land into a lucrative object of accumulation. Rapid urbanization in the erstwhile developing countries has made way for speculative investments in housing and infrastructure which is leading to an extension of the old cities and the transformation of small towns into large urban centres. Moreover, a regular occurrence is that urban policies in countries like India are made adaptable to gentrification and other zoning practices that lead to the dispossession of millions of people in the name of development and the recycling of their emptied plots for commercial purposes. Such accounts of financialization of space become even more relevant in the context of calculability if we recount Simmel's hypothesis that calculations proliferate with the rise in urbanization (Mitchell 2002, 80).

What is the connection between these two phenomena and the conception of the port as a site of logistical governance? Logistics, in this framework, can be defined as the process of bringing together the regimes of space-bound calculability and financialization. The logistical politics of space adheres to various geopolitical contentions appearing in different historical contexts. Historically, the will to govern the methods of calculation and financial impetuses leads to the constitution of certain semi-autonomous agencies or institutional apparatuses. These agencies share some properties with the government insofar as they channel or disperse assets and resources including human capital. At the same time, they are limited by the political will of government in terms of making autonomous choices regarding public policy. This produces a series of contradictions—especially in the context of public sector enterprises like ports or railways in India—which refuse to dissolve even when there is a specific logistical system at work. The issues of logistical governance, hence, have to take account of these contradictions which are immersed in the dynamic relationship between state and market.

Kolkata Port Trust as an Institutional Apparatus

The KoPT has been in charge of the management of the Kolkata Port since 1870. Founded by the colonial rulers of India, it was bestowed with the responsibility of expansion and management of the Calcutta Port at Kidderpore.[1] At the turn of the nineteenth century, the port in Kolkata saw a spurt in its traffic and augmentation of facilities. The export of coal, for example, rose from a mere 4282 tons in 1893–94 to 877,895 tons in 1898–99. Similarly, the export of food grains also shot up from 405 tons to over 200,000 tons in the same period. In 1914, at the onset of the First World War, the Kidderpore dock had 17 general cargo berths and 10 coal berths, with coal as a primary object of cargo movement. A chief import item in the second half of the nineteenth century was kerosene oil, and another important export item from Calcutta was tea, for which separate transit sheds and warehouses were installed along the river (Ray 1993, 157–158). After the Second World War, there was a period of slack in cargo traffic, and this continued until 1951. Then, during the Second Five Year Plan (1956–61), some recovery was made due to the government's decision to import iron, steel, and project cargoes. Under the same plan, the dock facilities were expanded with the purchase and replacement of cargo-handling equipment, cranes, railway tracks, diesel locomotives, and so on, and 113 gangs of secondary cargo and coal dock labourers, including 1500 temporary workers, were made permanent employees of the port (Ray 1993, 160).

There were special provisions for all the ports in the country in the different Five Year Plans of the Indian Government. After the depreciation of the port facilities during the Second World War, the First Five Year Plan (1951–56) put emphasis on the acquisition of 'new vessels like dredger, survey vessel, dock tug, anchor vessel, light vessel and launch' (Ray 1993, 161). The Second Five Year Plan, as we saw earlier, continued this scheme of reorganizing the facilities along with introducing formalization of port labour. The most important intervention in the Third Five Year Plan (1961–66) was the initiation of another dock at Haldia to assuage the pressure on the Calcutta Port. A further important decision, during this time, was to construct a barrage in the upstream of the river Hooghly, under the name of the Farakka project, to increase the headwater supply of the river to facilitate the draught for large vessels. This decision created a lot of controversy and geopolitical tension between India and Bangladesh because the later plans had major provisions for construction of the Haldia dock and the replacement of old technologies by developing container parks, installing computerized systems, and the modernization of the railway tracks.

If we have a closer look, we shall see that these provisions were the result of endless calculations and speculations about the geopolitical exclusivity of the port. Because it is a riverine port, Kolkata's narrow and tortuous approach is encumbered with numerous sand bars across the river Ganges. Thus, the port has the longest pilotage distance where the vessels have to shirk the sand bars and make intricate calculations about the height of the tides for easy draughting. Any detailed study of the movements of the ships will reveal the enactment of a complex interface between human skills and nonhuman predicaments, and it is crucial to understand that the nonhuman elements are not fixed components in a deterministic matrix of logistical governance. They also move, shift identities, and participate in international conflicts like the one that occurred between India and Bangladesh over the releasing of water from the river Padma through the Farakka barrage to allow Kolkata-bound vessels to draught comfortably.

The KoPT, which is under the directives of the Ministry of Shipping, Government of India, has two dock facilities under its control: the KDS and the HDC. Apart from the Board of Trustees, which is the apex decision-making body of the KoPT, there are a number of Principal Officers, headed by a chairperson from the Indian Administrative Service, who are in charge of the everyday activities at the port. The KDS has its own personnel responsible for financial and accounting activities, vigilance, marine engineering, hydraulic engineering, mechanical engineering, traffic, law, estate and materials (assets) management, and medical responsibilities. Similarly, the HDC has its own set of personnel in charge of marine, finance, traffic, and assets management (Kolkata Port Trust 2016b). Both ports under the KoPT have separate marine departments which are deployed to maintain the navigational channels through dredging and other measures. The traffic department handles all cargo operations including storage, loading, and unloading. The mechanical engineering department looks after the maintenance of cargo-handling equipment, vessels owned by the ports, electrical systems, lock gates, and locomotives. The port in Kolkata has a special research-oriented department, headed by the Chief Hydraulic Engineer, for studying river behaviour (Ray 1993, 214–15).

Apart from the officials in charge of different departments, a number of employees work at the ports on permanent and casual bases. As of 31 March 2014, the total number of employees at both Kolkata and Haldia ports was 7008, of whom 836 were Class-I and Class-II officers, 3936 were noncargo-handling Class-III and Class-IV staff, and 2237 were cargo-handling offshore and onshore workers (Kolkata Port Trust 2016b, 110). All dock workers at the Kolkata Port are covered by the Dock Workers' (Safety, Health and Welfare) Act 1986. The workers at the workshops are covered by the Factories Act 1948. The responsibility of ensuring safety at work, investigating accidents, and recommending remedies to health hazards is entrusted to a safety committee for each port which has as its members the port officials and users, representatives of the labour unions, and the Inspectorate of Dock Safety (Kolkata Port Trust 2016b, 52).

Most of the calculations regarding piloting, drafting, and dredging are done by the research staff under the Chief Hydraulic Engineer. The Kolkata Port is unique in two ways: one, as mentioned before, is that it has the longest pilotage distance between the sea and the port; and two, this navigational channel is abundant with sand bars created by deposits of silt in different sections of the river Hooghly. The navigational channel begins at the Sandheads in the Bay of Bengal, and the first anchorage point is located on the south-west side of Sagar Island, some 87 km north of the Sandheads. Most vessels come to Kolkata straight from the Sagar Anchorage through the Rangafala channel, moving zigzag in order to avoid the bars with the rise of the tide. Leaving from Kolkata towards the sea is even more complex, and any ship with a deep draught has to halt quite a few times between the port and the Sandheads, depending on the height and location of tides.

'Tide plays the most crucial role in pilotage to the Port,' a study of the Kolkata Port argues (Ray 1993, 184). A short description of the mechanisms that are required to calculate and predict the movement of tides is necessary to understand the nitty-gritty of logistical operations at the port:

> The rise of tide in the river varies from 4.2 m during neap tide to 6.5 m during spring tide. The bars have to be kept under constant watch to monitor the depth of water over them every day. At different crucial areas there are semaphores which show the depth over the bars at different times on rise and fall of the tide. Tidal semaphores (night) with acetylene flashing lights function at Mayapur, Hugli Point, Balari, Gangra and Sagar [all different

sand bars] which indicate in white, red and green colours the rise of tide at night. Tidal semaphores (day) are maintained at Kidderpore, Rajabagan, Akra, Mayapur, Hugli Point, Balari, Gangra and Sagar. The tidal position is shown by metre arm, decimetre arm and centimetre arm and by positioning a black ball which is kept high at rising tide and lowered down at low water and falling tide. (184–85)

Apart from these techniques, there are 'lighthouses, light vessels, lighted and unlighted buoys, track marks and towers on the shores to guide the pilots' (Ray 1993, 185). However, the job of piloting does not only involve technological brilliance but also need 'human' touch, as recounted by a pilot with enormous experience. Speaking of the changes brought about by new technologies, R. E. Mistry observes, 'Piloting has become less lonesome now' (1995, 111). In the past, when there was no such facility as VHF communication with the port authority or other ships, the pilot often had to take major decisions on the basis of his/her instincts and with assistance from absolutely random sources like the 'lone bobbing flare of a *mashal* (flame-torch) of a fisherman winding his way home at night' (111). With improvements in technology and hydraulic sciences, the unpredictability of the river has been brought under some control, but the 'hazards' of the Hooghly—a river famous for its 'Bars, Bores and Bends'—can only be mastered by individual skill and an undiminished love for the water body: 'Computers can work wonders but, for handling ships in the river Hughli [sic] we will still require quick judgement of a river pilot' (112).

This testimony is instructive in several ways. It tells us that the regime of calculability—which incorporates large-scale technical operations including measurement and analysis of tidal data, software applications, durable capital like vessels and buoys, and mechanical and civil engineering projects—is founded on a complex relationship between human skill and nonhuman obstacles. In this situation, the accuracy of systemic calculations (and the associated discourses of efficiency) does not exhaust the truth potential of the system because it is interspersed with stories of individual and collective skill, a nostalgic appraisal of certain institutions, and narratives of human virtue triumphing over the most obfuscating shortcomings. This narrative of a human surplus over machinic accuracy indicates the ingenuity with which a debilitating factor is converted into a positive sign of triumphant humanity. Other studies on the Kolkata Port have come up with similar stories where enterprise on the river is 'profoundly shaped by the actions and reputations of exemplary men, who are Kolkata Port Trust bureaucrats' (Bear 2015, 414). According to Laura Bear (2015), neoliberalism in India has created an environment of 'popularist speculation' where privatization, banking reforms, and reorganization of public debt have infused a culture of speculative investment in every individual and threatened the existence of bureaucratic institutions like the KoPT (408–23). In response, these institutions start to recoup by emphasizing the ingenuity of their respective speculative enterprises and grounding these moments of ingenuity in the essential function they play as harbingers of social relations.

This combination of speculative reasoning, managerial expertise, and social responsibility is explained by the editor of a volume of essays commemorating the 125th year of the Calcutta Port Trust in 1995 as an effect of the confluence of social, natural, and cultural functions (Chakraborty 1995, NA). While justifying the plan of the volume, Chakraborty describes the port as 'nothing more than an artifact' which can be put to many uses depending on the 'complex interplay of many social forces manifested as stakeholders.' These stakeholders are not necessarily human; they could be social motivations like the demands of the hinterland, natural factors like the tidal flow, or cultural determinants like the organization of the Port Trust. The question of skill also makes an appearance in this description. 'Anticipation of the motives of the society certainly calls for skill,' the editor informs, 'but one has to endeavour to acquire such a skill. If otherwise, the operators of this artifact (such as the Port) believes [sic] that it can handle the affairs as an autonomous entity, then it can only condemn itself by holding on to false promises' (Chakraborty 1995). This sums up the logistical framework within which the institutional apparatus of the KoPT has to operate: (1) it desires to *anticipate* the motives of the stakeholders; (2) it requires a set of *skills* to do so; (3) acquiring of that set of skills necessitates *interaction* with other stakeholders; and (4) without this interaction, a false sense of *autonomy* will arise. Therefore, the regime of calculability (which is space-bound and directed to govern uncertainties) cannot be based on the auto-corrective mechanism of liberal governmentality emulating the model of market autonomy. On the other hand, the sociality which is presumed by the interaction between different stakeholders is grounded in a politics of space that involves a series of unrelenting calculations.

The Question of Land

The politics of space in question, however, cannot evade another marker of our time—financialization of space in an urban context. Increasingly, the statements about inefficiency of the Kolkata Port are being linked with its locational disadvantage and the unutilized potential of the urban space under its control. It is often said that the Kolkata Port is dying because of difficulties in pilotage and drafting. Arvind Subramanian, the Chief Economic Advisor to the Indian Government, has recently advised the State Government of West Bengal to shut the ports in Kolkata and Haldia and use the vast tracts of land to 'create a global knowledge hub, tying into the state's well-known but underutilised human capital' (Business Standard 2015a). However, a look at its annual Administrative Report for 2013–14 indicates an ongoing process of recuperation with the Kolkata Port currently ranked third among all Indian major ports in terms of container traffic handling; second in terms of growth in handling both iron ore and fertilizer; and third in the handling of the raw materials for fertilizer. Also, Kolkata is ranked first in terms of the number of vessels handled during the financial year of 2013–14 (17.1 per cent of the total number of vessels handled in all Indian ports) (Kolkata Port Trust 2016b, 1). Numerous public-private partnership (PPP) projects are also underway including the development of berth facilities at the Haldia dock, betterment of transloading facilities at the Sandheads and its vicinity for midstream handling of dry bulk cargo, and development of a container terminal in Diamond Harbour. By the latest calculations, in the quarter of April–September 2015, a massive 19.62 per cent rise in cargo traffic was recorded over one year (from April to September 2014) under the Kolkata Port Trust (2015).

One reason of this upsurge is the increasing geo-spatial importance of the Kolkata Port in South-East Asia. With the realization of the New Silk Route in near future, the port in Kolkata is destined to become a strategic nodal point in an international trade network alongside ports in neighbouring countries like Myanmar and Bangladesh. The Government of India has also started to take notice of its geopolitical potential and, accordingly, is strategizing its 'modernization' through a major port linking Chennai (India) with Yangon (Myanmar) and Chittagong (Bangladesh) in its latest scheme, titled 'Sagarmala,' to improve maritime trade (Ministry of Shipping 2015). The modernization drive will focus on development of efficient coastal transport networks, promotion of port-based special economic zones (SEZ) and ancillary industries, and the enhancement of tourism and aestheticization opportunities. The Union Shipping Minister, Nitin Gadkari, has recently revealed that the total investment in this project will exceed INR 70 billion (*The Economic Times* 2015).

One of the crucial features of the Sagarmala project is its insistence on utilizing the space in and around the docks by creating investment opportunities in the land under the ownership of the port authorities like KoPT. Being the city of Kolkata's largest owner of land, the KoPT thus emerges at the centre of a hotbed of land speculation, rent extraction, and financialization of space (Ray 1993, 206). Right now, the port authority owns different-sized parcels of land scattered all over the city, most of which are leased out for various residential and commercial purposes. It also extracts rent from the numerous warehouses it owns in Kolkata: the Strand warehouses, the Armenian Ghat Warehouse, the Canning Warehouse, the Clive Warehouse, and so on. The rent income of the KoPT is yet to become a major source of revenue for the port, but the annual Administrative Report (2013–14) shows a small increase in rent and premium on leased land (INR 24.1 million) from the previous year. However, as newspaper reports show, KoPT has become quite alert to the potential of remodelling these land parcels into more economically viable spaces of rent extraction and is trying to recalibrate the older rates and schedules. It is expecting a 14 per cent increase in revenue from leasing its land in the fiscal year of 2015–16 and is considering many other options in land speculation and utilization (Business-Standard 2015b).

The Ministry of Shipping has been issuing policy guidelines for the use of land by the major port trusts since the passing of the Major Port Trusts Act 1963. According to the Act, the lease of any immovable property, including of land to private parties, must not exceed 30 years without prior approval of the Central Government. In 2012, a draft policy for land management by the major ports was proposed by the Ministry of Shipping. It was finalized in 2014 after interministerial consultations and interventions by the Indian Ports Association.[2] The main objectives of this policy are to ensure optimization of the use of land resources and transparency of land-related transactions (Kolkata Port Trust 2014). However, it also states that separate policy needs to be formed for land holdings in the township areas of Kolkata and Mumbai, two of the most heavily populated urban centres in India.

Accordingly, a document regarding 'Land Use Plan/Zoning' of the estate of KoPT in Kolkata (under the jurisdiction of the KDS) was prepared and uploaded onto the Port Trust website in January 2016. It invited comments and suggestions from the citizens of India (Kolkata Port Trust 2016c). In this scheme, the land parcels are distributed among 33 zones specifying the location, prevailing land use patterns, and recommended changes in such patterns. To give an example, Cossipore (Zone 1) which now has a concentration of residential buildings and business and educational establishments should in coming years become a tourist hub with open riverfront spaces, plaza, recreational centres, and mercantile storage options. While most of the zones are recorded in the document to have a similar concentration of residential and business housing, the proposed land use plans differ according to locational specificities. Whereas Cossipore and the land adjacent to Circular Canal from Chitpur in North Kolkata to Tolly's Nullah in the South (Zone 2) are recommended for landscaping, tourism activities, parks, and other recreational facilities, the land close to the dock in Garden Reach (Zone 3) is suggested to be preserved for mercantile activities, the extension of existing industrial establishments, storage, dry docking, boat and vessel repair, cargo-handling, port-related allied facilities, and jetties (Kolkata Port Trust 2016c).

The KoPT (2016c) document also contains remarks from the issuing authority about each zone. Most start with a prosaic declaration: 'The proposed land use is largely in conformity with LUDCP [Land Use and Development Control Plan] of KMDA [Kolkata Metropolitan Development Authority].' However, in a few cases, we find interesting observations. In Chetla (Zone 22), the document has recorded existence of small workshops along with residential buildings. Noting that these workshops do not conform to KMDA's land use policy, the document opines, 'Considering reality, the existing workshops may continue with permission of KMDA.' Similarly, for the land between Nityadhan Mukherjee Road and Jagat Banerjee Ghat Road and the adjacent area (Zone 24), the recommendations consist of leases to assembly, storage, business, and mercantile establishments, and not residential buildings, but '[b]ecause of high potential of the area for use as residential purpose in future, the Land Use may be reviewed after 10 years to explore whether the same may be confined to residential buildings only' (Kolkata Port Trust 2016c).

The document and the remarks therein are important for two reasons: one, they indicate the Port Trust's eagerness to financialize the land parcels under its control in accordance with the reforms suggested by the central government's policy guidelines. But more crucially, it points to the negotiations that KoPT has to undergo with other government agencies like KMDA in order to emerge as an important player in the urban land market. This precondition is already confirmed by the government's guideline which has recommended a separate policy for the urban land under KoPT. In preparing and circulating the document, the KoPT also elucidates a changing dynamic of stakeholding in which urban development authorities are now recognized as legitimate stakeholders in the operations of the Port Trust at the level of logistical governance.

In the age of deregulation, the ports are required to be financially self-sustaining. While a major source of this self-sustenance has to be the hitherto less-explored area of urban land speculation, that too has to happen within a seemingly transparent field of public discourse. As well as uploading the proposed land use plan to its website for public review, the KoPT has also published a list of market rates of the different zones of land along with 'offered tender rates,' as required by the government's policy guideline (Kolkata Port Trust 2016d). It is difficult to estimate how much of this desire to take cognizance of public scrutiny will translate into actual results, but it definitely highlights the exclusivity of logistical governance of the port as a public sector enterprise. On the other hand, how this public is constituted and what roles and responsibilities it entails need to be explored carefully. The spatial overhaul prescribed in the document is likely to dispossess many if forceful land-grabbing is allowed to take place. Will those at risk of dispossession have any say in the formation of the land use policy? Most probably not. Meanwhile, the process of marking the territories and driving out the illegal squatters has started, as is clearly evidenced in a recent squabble between KoPT and a film production company which continued to run its business at an 80,400 square feet plot in the Hyde Road Extension despite the expiry of its lease and the port authority's refusal to renew it (*Times of India* 2015).

So far I have tried to underscore the linkages between calculations governed by spatial considerations and speculations related to space-making exercises so that the material foundations of logistical governance come to the surface. What is even more interesting in this context is the fact that KoPT remains a public sector enterprise with thousands of permanent staff and millions of dollars in built-in assets—a typical case in many Asian countries. The connections between various forms of calculation about the details of pilotage and the drafting and modalities of financialization of space by reforms in rent structure and revaluation of land holdings cannot be addressed if we do not consider the governmental apparatuses that are in operation here. But another point needs to be considered in this context. The broader aspect of financialization encompasses a domain of calculability which tends to transcend spatial coordinates in the first instance. In the case of the port, for example, there are functions, motivations, aspirations, and institutions which are not exclusively spatially organized; in fact, in tandem with the global financial order, another regime of calculability dominates the policy decisions and public discourses, for example, the calculations that refer to revenue and expenditure of the port system, valuation and depreciation of human and nonhuman assets, risk assessment, and insurance technologies. Often these calculations expose the contradictions between different elements in the government, between policy recommendations and the 'autonomous' working of institutional apparatuses like the KoPT. For example, when there is a strong emphasis by policymakers on the liquidation of port assets and investments to create knowledge hubs, the port authority insists on carrying out its operations as before. However, the inclusion of new stakeholders, not only through networks with other government agencies like KMDA, but also with the increasing participation of international funding agencies like the Asian Development Bank (ADB) for facilitating trade in South and South-East Asia (ADB 2015), draws our attention to another interactive paradigm which anticipates the world of logistical governance. In the final section, I shall dwell on the specificity of this 'logistical world' in connection with the geopolitics of infrastructural connectivity in South and South-East Asia.

The Paradigm of Asian Connectivity

The rising interest in Kolkata Port coincides with a growing recognition of its locational advantage in the schemes proposed under India's Look East Policy (renamed as the Act East Policy in 2014), whose main thrust has been to forge sustainable political and economic relationships with its neighbouring countries in South-East Asia so that India can compete with China as a regional power, especially in the context of Asia's emergence as a leader of globalization following the economic meltdown in the West. This attitude must be analyzed in terms of India's changing relationship with Asian power blocs and regional conglomerates—such as the Association of South-East Asian Nations (ASEAN) and the Bay of Bengal Initiative for Multisectoral Technical and Economic Cooperation (BIMSTEC)—while keeping in view its need for improved trade connectivity in the region (Shrivastava 2005; ASEAN 2015; Lee 2015). Hence, a report by the ADB recommends developing a coordinated regional road development programme, an upgrade of border link roads, and the construction of deep-water ports efforts in order to solve the problems of restricted draught and limited navigation of large vessels in ports in the northern part of the Bay of Bengal (Chittagong, Kolkata, and Haldia) (ADB 2008, ix–xv). With financial and technical support from the ADB and the World Bank, many projects are already underway to overcome various bottlenecks in transport infrastructure. These include the building of cross-border infrastructure between India and Thailand, the construction of port-based SEZs in Myanmar, and the planning of an India-Myanmar-Thailand Trilateral Highway linking Moreh in India with MaeSot in Thailand (De 2016, 2).

The revamping of logistical infrastructure in South and South-East Asia is crucial to the work of the ADB. According to its website, 80 per cent of its lending to member countries is concentrated on infrastructure, education, environment, regional cooperation and integration, and financial reforms (ADB 2016). Most of the infrastructure money goes into funding the improvement of transportation, and the ADB has already established a programme for its developing member countries (DMC) to promote the concept of 'sustainable transport initiatives' that is supposed to allow 'basic access and development needs of individuals, companies, and society to be met safely and in a manner consistent with human health' (ADB 2010, 4). This concept is coterminous with ADB's vision of Asia as a growing economic region where obstacles in logistical infrastructure will be mitigated by installing a strong network of regional integration. While the ADB acknowledges that '[p]hysical connectivity is the bedrock of many economic cooperation and integration efforts' (11), it also maintains that the 'hardware' of physical connectivity across the region—construction of roads, bridges, ports, rail lines—must concur with its 'software,' that is, legal and regulatory frameworks and systems of customs clearance. In that sense, regional cooperation will require uniform regulatory and fiscal frameworks across borders including the 'harmonization of regulations, procedures, and standards' (11). This combination of hardware and software pertains to the concept of Asia itself as an integrated infrastructural project. Thus, multiple publications by the ADB and the Asian Development Bank Institute (ADBI)—the research wing of the ADB which has been operating from Tokyo since 1997—propose the building of an infrastructurally 'seamless' Asia (ADB and ADBI 2009; De and Iyengar 2014; Plummer et al. 2016). These studies agree that 'the time is ripe for research on cross-regional integration' and that this helps obviate any impending economic crisis (Yoshino 2016, ix).

The time is also ripe for initiating massive infrastructural activities to match these studies. Kolkata—the city itself and its port—has attracted the attention of the designers of a 'seamless Asia' because of its relatively central location in the grand scheme of things in South and South-East Asia. So much so that it could be projected as a logistical hub in the region. The studies on a connected Asia cannot ignore the reality of its centrality with respect to various logistical chains that are needed to facilitate the cross-border movement of labour, capital, and commodities. As one study on land connectivity in South and South-East Asia observes, 'In South Asia all corridors originate from Kolkata and Chittagong ports in the Gulf of Bengal' (Gautrin 2016, 38). The study further mentions 'five possible road corridors for South Asia,' of which three start in Kolkata: the Kolkata-Chicken's Neck (via Siliguri) corridor (Manipur), the Kolkata-Bangladesh corridor, and the Kolkata-Chicken's Neck corridor (Mizoram) (39).

Similarly, due to the possibility of its being at the centre of numerous road and rail corridors—as well as its depleted yet diverse infrastructural facilities and the richness and dexterity of its human capital—the Kolkata Port is acquiring a special place in the paradigm of Asian connectivity, especially in the Bay of Bengal region. For example, a glaring example of PPP in the Asian connectivity paradigm is the Kaladan Multi-modal Transit Transport Project which is supposed to connect the Kolkata Port to Sittwe port in Myanmar (Kabir and Ahamad 2015, 222). A deep-water port will be built jointly at the Sagar Island—150 km in the downstream of Kolkata—by the KoPT and the Government of West Bengal (Live Mint 2015). This port, along with the one in Haldia, will operate as part of the integrated port system under the KoPT. Apart from taking care of the large vessels which cannot anchor at the Kolkata Port, due to the restricted draught, it will also operate as a military base for Indian surveillance of the Bay of Bengal region (*Times of India* 2016).

It is debatable how much of this will come to fruition; what is palpable is the urgency with which plans of regional integration, schemes of infrastructural development, strategies of military concentration, and designs for removing obstacles to the movement of capital are coming together to produce a logistical world.

This logistical world is also 'patterned' by the postcolonial invention of area studies and the regionalism that it promotes. As Sandro Mezzadra and Brett Neilson argue, the 'rise of area studies...involved an effort to bestow a sense of scientific authority and objectivity on the division of the world into more or less boundable areas, supposedly united by social and cultural features and understood as comparable and thus separable entities' (Mezzadra and Neilson 2013, 42). During the Cold War, these areas played their respective roles as 'lackeys' of the American or Soviet camp, or despite efforts to remain unattached, like India, but finally leaned on one or the other. In the early 1990s, the same regional blocs transformed themselves into cogs of a huge infrastructural machine whose formation and sustenance is coterminous with increasing physical and virtual connectivity between the same regions. This conception of the continent—breakable into regional conglomerates but also presentable as a seamless unity when needed—is impossible without addressing the question of infrastructure as its organizing principle. Similarly, infrastructure in this context is defined as a political entity whose blueprint is being drawn in the interplay of global capital and local aspirations.

CONCLUSION

It would be a mistake to conclude that this geopolitics of infrastructural connectivity is explanatory for the growing popularity of the Kolkata Port. In a way, while the paradigm of Asian connectivity does explain reasons for the port's so-called revival, that is not all. In the beginning, I described the KoPT as an institutional apparatus—a dynamic site of calculations and speculations which facilitates interactions between infrastructure and logistics. At the same time, it plays the role of a 'model' of accumulation that neoliberalism proffers—the kind of accumulation evinced in the shift from the auto-corrective mechanism of liberal governmentality (modelled after the market) to the interventionist role of the nation-state in controlling informality in the economy. I use the term 'model' in the specific sense that is deployed by Mary Morgan and Marcel Boumans (2004) in their study of the methodology of economic analysis in the 1940s (369–401). According to them, the introduction of three-dimensional physical models to explain the workings of a Keynesian macroeconomic system is an important moment in the history of economic analysis. This marked a shift from the earlier two-dimensional (diagrammatic) models which were mainly epistemic exercises in abstraction and added a dialectic of constraints and commitment to the economist's list of concerns: constraints due to the physicality of the model—its concrete existence in actual coordinates of space and time; its size, shape, and angles—and commitment to the theory that it is supposed to represent in spite of these constraints.

If we imagine the KoPT as a physical model of neoliberal accumulation, we discover a similar framework of constraints and commitment in operation. There are undeniable constraints in the form of built-in capital, land holdings, employees, stakeholders, and natural impediments. There is also a strong commitment to the ideas of deregulation and disinvestment on part of the government. The function of a model is to explain the world by finding a perfect balance between constraints and commitment. The constraints allude to the obstacles in reality; the commitment reminds of a possibility of overcoming them. The port is a perfect example of such negotiations where calculations to overcome the constraints lead to speculations sanctioned by a commitment to neoliberalism. The partial autonomy of the KoPT also makes sense in this framework, as Morgan and Morrison (1999) observe, as the model must enjoy some autonomy from both theoretical commitments and realistic constraints for better, more accurate analysis (10–37). The KoPT's autonomy ensures that the balancing act, which is apparently free from any kind of favouritism, yields optimum skill and efficiency for running the port.

This framework is really useful to situate the Kolkata Port within a discursive exposition of neoliberalism. The continued discussions on connected Asia, and the implications and influences of this debate on the port's future, make one wonder if any such infrastructural endeavour requires internalization of some sort of structural ambivalence. The definition of 'sustainable transport' offered by the ADB mentions the necessity of reducing 'use of land and emissions, waste and noise' (ADB 2010). Waste and noises are inevitable in any infrastructural project—both literally and metaphorically—in terms of 'environmental' challenges and handling of disputes and resistance, respectively. Ironically, it becomes more difficult day-by-day as, unlike the olden days of unconcerned progress of imperial variety, the ADB and other suchlike organizations are also keen to alleviate poverty and guarantee socioeconomic improvement, and hence, cannot suppress all the noises emerging out of the clamour for sheer developmentalism.

Undoubtedly, this new order of imperialism is also geared to transform the social infrastructure which will embody and preserve the physical and virtual infrastructure. The 'social' in this scenario is not an unperturbed outside which provides a mere context to all that is happening. It is a continuous process of negotiations and renegotiations between different agents of growth, interplay of interests and concerns over the number and identity of various stakeholders, and the recognition and reorganization of the patterns of the world(s) we inhabit. The Kolkata Port, with all its limitations and lamentations, is a model of such instances of socialization of infrastructure and deserves our attention for a more cogent understanding of our time.

Notes

1. Currently, KoPT handles both the Kolkata Dock System (KDS) in Kolkata and the Haldia Dock Complex (HDC) in Haldia—The Calcutta Port Rules (1994) specify that the 'docks' under KoPT include 'Kidderpore Docks, Netaji Subhash Dock, Garden Reach Jetties, [and] Haldia Oil Jetties' (Ministry of Surface Transport 1995, Chapter 1, Article 12).
2. Indian Ports Association is a 'full-fledged professional body [which] renders Consultancy services on variety of subjects related to Port Development, improvement of Operational Efficiency and various issues directly involved in the overall Port Management' (Indian Port Association 2015).

References

ADB and ADBI. 2009. *Infrastructure for a Seamless Asia*. Tokyo: Asian Development Bank Institute.
ADB. 2008. *The Technical Assistance Consultant's Report: Bay of Bengal Initiative for Multi Sectoral Technical and Economic Cooperation Transport Infrastructure and Logistics Study*. Accessed 17 August 2016. http://www.adb.org/sites/default/files/project-document/65311/38396-01-reg-tacr.pdf
ADB. 2010. *Sustainable Transport Initiative: Operational Plan*. Manila: Asian Development Bank. Accessed 18 August 2016. http://www. adb. org/sites/default/files/institutional-document/31315/sustainable-transport-initiative. pdf
ADB. 2015. *Connecting South Asia and South-East Asia*. 2015. Accessed 11 November 2015. http://www.adb.org/sites/default/files/publication/159083/adbi-connecting-south-asia-South-East-asia.pdf
ADB. 2016. 'Policies and Strategies.' Accessed 18 August 2016. http://www.adb.org/about/policies-and-strategies
ASEAN. 2015. 'Chairman's Statement of the 13th ASEAN-India Summit.' Accessed 9 August 2016. http://asean.org/wp-content/uploads/images/2015/November/27th-summit/statement/FinalChairmans%-20Statement%20of%2013th%20ASEAN%20INDIA%20Summit.pdf
Bear, Laura. 2015. 'Capitalist Divination: Popularist Speculators and Technologies of Imagination on the Hooghly River.' *Comparative Studies of South Asia, Africa and the Middle East* 35, no. 3: 408–23.

Business Standard. 2015a. 'CEA says Kolkata port Great Example of Infrastructure Draining Public Money.' 28 December. Accessed 5 January 2016. http://www.business-standard.com/article/economy-policy/cea-says-Kolkata-port-great-example-of-infrastructure-draining-public-money-115122800032_1.html

Business Standard. 2015b. 'Kolkata Port Eyes 14% Revenue Increase in Land Leases.' 18 December. Accessed 5 January 2016. http://www.business-standard.com/article/companies/Kolkata-port-eyes-14-revenue-increase-in-land-leases-115121800605_1.html

Chakraborty, Satyesh C. ed. 1995. *Port of Calcutta: 125 Years.* Calcutta: Calcutta Port Trust.

De, Prabir and Kavita Iyengar, eds. 2014. *Developing Economic Corridors in South Asia.* Manila: Asian Development Bank.

De, Prabir. 2016. 'Strengthening BIMSTEC Integration: The New Agenda.' *BIMSTEC Newsletter* 72, no. 1: 1–3.

Gautrin, Jean-François. 2016. 'Land-Based Cross-border Transport Infrastructure.' In *Connecting Asia: Infrastructure for Integrating South and South-East Asia,* edited by Michael G. Plummer, Peter J. Morgan, and Ganeshan Wignaraja, 37–69. Cheltenham and Northampton, MA: Edward Elgar.

Harvey, David. 2009. *Social Justice and the City.* Athens and London: University of Georgia Press.

Indian Port Association. 2015. 'About Us.' Accessed 6 January 2015. http://ipa.nic.in/index1.cshtml?lsid=13

Kabir, Mohammad Humayun, and Amamah Ahmad. 2015. 'The Bay of Bengal: Next Theatre for Strategic Power Play in Asia.' Croatian International Relations Review, no. 72: 199–239.

Kolkata Port Trust. 2015. Administrative Report, 2013–14. Kolkata: Kolkata Port Trust.

Kolkata Port Trust. 2014. 'Policy Guidelines for Land Management by Major Ports, 2014.' Accessed 5 January 2016. http://www.Kolkataporttrust.gov.in/showfile.php?layout=1&lang=1&level=1&sublinkid=1786&lid=1507

Kolkata Port Trust. 2016a. 'Salient Features.' Accessed 18 August 2016. http://www.Kolkataporttrust.gov.in/index1.php?layout=1&lang=1&level=1&sublinkid=103&lid=23

Kolkata Port Trust. 2016b. 'Administrative Report, 2013–14.' Accessed 22 August 2016. http://Kolkataporttrust.gov.in/showfile.php?layout=1&lang=1&level=2&sublinkid=1835&lid=1552

Kolkata Port Trust. 2016c. "The Proposed Land Use Plan/Zoning, 2015 of KoPT Estate Under KDS.' Accessed 22 August 2016. http://www.Kolkataporttrust.gov.in/showfile.php?layout=2&lid=1572

Kolkata Port Trust. 2016d. 'Market Rate of Different Zones of Land According to SoR (Long Term Lease).' Accessed 22 August 2016. http://www.Kolkataporttrust.gov.in/showfile.php?layout=1&lid=1675

Lee, Lavina. 2015. 'India as a Nation of Consequence in Asia: The Potential and Limitations of India's "Act East" Policy.' *The Journal of East Asian Affairs* 29, no. 2: 67–104.

Live Mint. 2015. 'Kolkata Port Trust, West Bengal Govt to Sign JV for New Port,' 7 January. Accessed 18 August 2016. http://www.livemint.com/Industry/rwWeoFww5eWhg1N6Z5lb9I/Kolkata-Port-Trust-West-Bengal-govt-to-sign-JV-for-new-port.html

Mezzadra, Sandro and Brett Neilson. 2013. *Border as Method, or, the Multiplication of Labor*. Durham and London: Duke University Press.

Ministry of Shipping. 2015. 'Concept Note on Sagarmala Project: Working Paper.' Accessed 22 August 2016. http://www.ipa.nic.in/Conceptnote.pdf

Mistry, R. E. 1995. 'Memories of Pilotage.' In *Port of Calcutta: 125 Years*, edited by Satyesh C. Chakraborty, 110–112. Calcutta: Calcutta Port Trust.

Mitchell, Timothy. 2002. *Rule of Experts: Egypt, Techno-Politics, Modernity*. Berkley: University of California Press.

Morgan, Mary, and Margaret Morrison, 1999. 'Models as Mediating Instruments.' In *Models as Mediators: Perspectives on Natural and Social Science*, edited by Mary S. Morgan and Margaret Morrison, 10–37. Cambridge: Cambridge University Press.

Morgan, Mary S. and Marcel Boumans. 2004. 'Secrets Hidden by Two-Dimensionality: The Economy as a Hydraulic Machine.' In *Models: The Third Dimension of Science*, edited by Soraya Chadarevian and Nick Hopwood, 369–401. Stanford, California, Stanford University Press.

Plummer, Michael G., Peter J. Morgan, and Ganeshan Wignaraja, eds. 2016. *Connecting Asia: Infrastructure for Integrating South and South-East Asia*. Cheltenham and Northampton, MA: Edward Elgar.

Ray, Animesh. *Maritime India: Ports and Ships*. Delhi: Pearl Publishers, 1993.

Shrivastava, Smita. 2005. 'BIMSTEC: Political Implications for India.' *The Indian Journal of Political Science* 66, no. 4: 973–88.

The Economic Times. 2015. 'Sagarmala Project: Government to Spend Rs 70,000 crore [70 billion] on 12 Major Ports, Says Nitin Gadkari.' 6 October.

Times of India. 2015. 'Venkatesh Films to Vacate Port Land by Nov 16: Calcutta HC.' 16 October. Accessed 11 November 2015. http://timesofIndia.Indiatimes.com/city/Kolkata/Venkatesh-Films-to-Vacate-port-land-by-Nov-16-Calcutta-HC/articleshow/49398573.cms

Times of India. 2016. 'NHAI to construct bridge to Sagar Island.' 28 April 2016. Accessed 18 August 2016. http://timesofIndia.Indiatimes.com/city/Kolkata/NHAI-to-construct-bridge-to-Sagar-Island/articleshow/52018921.cms

Yoshino, Naoyuki. 2016. 'Foreword.' In *Connecting Asia: Infrastructure for Integrating South and Southeast Asia*, edited by Michael G. Plummer, Peter J. Morgan, and Ganeshan Wignaraja, ix–xi. Cheltenham and Northampton, MA: Edward Elgar.

CHAPTER 4

Ports and Crime

Paula Banerjee

Ports are habitually at the intersection of security and insecurity, crime and crime control, petty theft and transnational offence; and yet, while port areas and harbours often witness unusual violence, life within them goes on as usual. Even though national networks and transnational flows converge in ports, until recently, they were not considered to be prime sites for the study of border criminology. Crime and everyday violence have seemed like life as usual in ports, and it is only since 11 September 2001 that their full potential for crime has dawned on the brokers of security. Why did people suddenly realize that ports were endangered by a new wave of crime? Ashley Roach gives the following reasons:

1. Approximately 90 percent of the world's cargo moves by ship.
2. Globally, each year over 48 million full cargo containers move between major seaports; overall some 200 million full and empty containers are transported between the world's seaports.
3. Each year, more than six million containers are off loaded at US seaports.
4. Almost half of the incoming US trade (by value) arrives by ship.
5. More than US $ 1.2 trillion in imported goods passed through the United States' 301 ports of entry in 2001. (Roach 2003, 342)

P. Banerjee (✉)
Calcutta Research Group, Kolkata, West Bengal, India

© The Author(s) 2018
B. Neilson et al. (eds.), *Logistical Asia*,
https://doi.org/10.1007/978-981-10-8333-4_4

The mixture of insecurity and economy makes ports unique spaces. The business of ports makes it inevitable that transnational flows of goods, people, diseases, and fear no longer remain foreign and, thus, at a distance. Ports make national space transnational, and all that is considered polluting cannot be kept separate from so-called safe national spaces. Ports are where containers deliver goods that could potentially pose a danger or threat. As one expert on criminology points out, 'cargo shipping itself contains insecurity to the extent that containers are ideal boxed spaces to transport illegal drugs and immigrants' (Eski 2011, 416). This makes ports an epicentre of insecurity. Even when there are hardly any threats, the spectre of containers bringing in weapons, drugs, people, arms, and other toxic products makes them spaces where such insecurity also makes violence routine. It is known that all containers are never checked, because that is business impossibility, so the potential for illicit trade remains, and the only people who have the power to control it are also those with the potential to become players in it. This makes the brokers of security in port areas also agents of insecurity.

When the question of flows—whether of goods, cash, drugs, or people—enters, there is potential for enormous profit and the keepers of security can also become ushers of insecurity, and they often emerge as prime suspects in this cusp between legality and illegality, legitimate and criminal activity. When the brokers of security become agents of insecurity, then governance to control crime becomes governance by crime.

Routine or Rarest of the Rare?

Case 1

On 9 March 2016, the Chairman of the Kolkata Port Trust, R. P. S. Kahlon, was arrested after he allegedly accepted a bribe of INR 2 million in a five-star hotel in Kolkata. At the time of his arrest, the police alleged that this was not the first time that Kahlon 'had come under the scanner over the past few months' (Chaudhuri 2016a, 20). The bribe was given by a Jagtap Deoji who was involved in the container business and handled containers in the Kolkata Dock System. Initially, the two men were remanded in police custody until 17 March 2016, after which they were taken to jail. Predictably, Kahlon's lawyers were crying foul play, but Kahlon was relieved of his job as Chairman of the Kolkata Port Trust on 23 March 2016.

While investigating this case, I found two distinctly different versions of who Kahlon was. Some described him as a polite but strict officer who, as the Chairman of the Kolkata Port Trust, took a number of 'fearless steps,' of which one was to foreclose and remove the studio of a well-known film producer, Srikant Mohta, from the Brace Bridge area. In this case, the Calcutta High Court subsequently upheld the decision taken by the Kolkata Port Trust and, since Srikant Mohta was close to West Bengal Chief Minister (CM) Mamata Banerjee, this group was of the opinion that Kahlon was a victim of foul play and his arrest was politically motivated. Kahlon, they said, was a nice man who had been victimized because of his opposition to Mohta, and that is why the state police did not involve the CBI before taking action against him. It also explains, they argued, why Kahlon was kept away from the press after his arrest (*Bartaman* 2016, 1).

Others, however, were closer to the police in their opinion of Kahlon. They believed he had been systematically abusing his office for personal gain and also alleged that, after becoming the Chairman of the Kolkata Port Trust, he used to frequent the five-star hotel where he was ultimately arrested. They also alleged that this was where Kahlon met with different business people. The questions they wanted answers to were: Why did Kahlon meet with so many business people? Did he frequently meet with them to extort money? What did he do with the money? Was it sent abroad through the *hawala* money laundering channels? (*Aajkal* 2016, 8). Another newspaper speculated that Kahlon was laundering money to Australia and South-East Asia and was under the radar of an Australian investigative agency: 'According to the intelligence shared by the foreign agency, funds were transferred from India to Australia through three means – online transfer through bank accounts registered against fictitious companies, *hawala* and human carriers. "Kahlon's money was transferred through at least one human carrier," an officer said' (Chaudhuri 2016b, 21).

Kahlon's case did not remain a simple case of one individual's corruption and/or money laundering. It assumed a larger significance beyond the port and hinted at 'other' connections that were central to the political mews of the state. When it was tagged with issues pertaining to other countries, in South-East Asia and Australia, the questions asked took on a different character. From being a commentary on the character of one man, the case evolved into one of conspiracy to undermine the economic security of the nation. Internal politics, electoral politics, and national security all became enmeshed and what had been a peripheral port crime transformed into a debate that was much more central to national security. However, the crime was such that it fitted in well with port logistics. Connection with South-East Asia is what the port is meant for. Sending money there logistically follows the route taken by legitimate business. But when crime is involved, port logistical concerns can become a central issue in state politics.

Case 2

On 12 February 2013, an incident took place in which members of the student wing of the Congress Party clashed with the student wing of the All India Trinamool Congress (TMC) that was in power in the state of West Bengal. It was the day for filing nomination papers, in Harimahan Ghosh College in Garden Reach (Kolkata Port) area, for the upcoming students' union elections. This is an extremely sensitive area because it is part of the dock area, and it was said that the Congress goons were led by a man named Mukhtar and the TMC by Mohammad Iqbal, alias Munna, who was a Trinamool Councillor close to the then Urban Development Minister, Firhad Hakim. In the clash, four students and a police sub-inspector (SI), Tapas Chowdhury, were killed.

When the incident first came to people's knowledge, it was reported that Mukhtar's men killed the SI. According to a news report, 'a group of 40-odd miscreants, led by a Congress worker, Mukhtar, arrived at the college and tried to force their way onto the premises. They used crude bombs to trigger panic on the premises. Urban development and municipal affairs minister Firhad Hakim alleged that when sub-inspector (special branch) Tapas Chowdhury chased Mukhtar, he opened fire and grievously injured the officer' (Bhabani 2013). Then the drama unfolded quickly as the case was taken over by the Crime Investigation Department (CID) of the local police. A few days later, news began appearing that it was not Mukhtar's men but Iqbal's men who killed the SI. Allegedly it was Firhad Hakim's confusing accounts of the event that baffled the police in the beginning, but on further investigation, they found that the trigger that killed the SI was in the hands of a follower of Iqbal. Condemnation came in fast. The Governor of West Bengal at that time was K. R. Narayanan who criticized Hakim's efforts to shield the TMC councillor: '"[Hakim] has no business to do that," Narayanan said when asked to comment on the state Urban Development minister Firhad Hakim's alleged bid to shield TMC councillor Iqbal from arrest for gunning down Special Branch sub-inspector Tapas Chowdhury on February 12' (*Times Now* 2013b). Even the Chief Minister distanced herself from the sensitive matter.

Immediately on learning that he was the prime suspect, Iqbal absconded. It was said that during the first three days of the killings, Iqbal was very much present in the city. But after he escaped, police sources traced him to Aligarh and the CID followed the trace. In Aligarh, however, police were stopped from performing their duties by Iqbal's men, and the CID was forced to arrest some of them for obstructing police work. By that time, Iqbal had disappeared once again. By the time he was found and arrested in a district in Bihar, it was more than 23 days after the killing of the SI (*Express News Service* Kolkata 2013). Meanwhile, the CM had promised all kinds of support to the dead SI's family including a clerical job for his daughter and provision for his son's education (*Times Now* 2013a).

There were two First Information Reports (FIRs) lodged over the entire incident and both were by men working in the Rapid Action Force (RAF). The first FIR was in the name of RAF inspector Milan Kumar Dam who wrote that the primary suspects were 'a fair looking, bald headed, aged man whom everybody was calling as Chairman Saheb' and another man 'aged about 25 years, medium height and medium complexion.' In the second FIR, by Garden Reach Officer in Charge (OC) Nazrul Islam, there was a more direct reference to Iqbal alias Munna. Islam clearly states that the person responsible was called 'Chairman Saheb alias Munna.' He also makes Mukhtar responsible for rioting in the area (Sen 2013). The killing of the SI soon took on a stronger political hue when it was alleged in the media that Firhad Hakim was doing his best to save Iqbal to the extent of obstructing justice. Once again this illustrated how the effects of a port crime could not be contained in the port but could expand to affect the politics of the entire region. The political rumblings intensified still further when the Police Commissioner, R. K. Pachnanda, who was considered capable and honest, was suddenly removed from his post.

The CID 'submitted a charge sheet naming Trinamul councillor Mohammad "Munna" Iqbal as the mastermind in police officer Tapas Chowdhury's murder, [and] the case against him was built around 122 witness accounts' (*The Telegraph* 2013). When the charge sheet was submitted in the Alipur court 57 days after Iqbal was arrested, it immediately started a debate about whether it would withstand court scrutiny. Some said that the grounds were so flimsy that it would actually aid Iqbal to walk away a free man. One anonymous respondent, who was himself in the police, explained that it is not at all necessary to visibly pull the trigger to be charged with second-degree murder. Other police sources said it was evident that the state government did not want to influence the investigations because the police were able to slap murder charges on Iqbal and his allies.

An incident that started as a contest between two political parties to claim a single college soon snowballed into a riot in the port area between vigilante groups, party goons, and the police. Then like an all-consuming forest fire it became a political debate over the ethics of the ruling party. What was yet to come was its transformation into Islamic terrorism in certain discourses. Three days after the killings, an article on *Covert Wires* (2013) went viral. In it, the author made the allegation that Indian democracy is all about appeasing the minorities by attacking and chastising the Hindus or the majority community. By now, the incident had been deemed to be part of worldwide Islamic terrorism, at least in certain discourses. A crime in the port area of Kolkata spread its tentacles and soon became a spectre that could not be contained within the ward borders.

Case 3

The third case is probably the most ignominious, and best known, of reported port crimes in the last 60 years. In it, the social world of the port is ambiguous, largely due to the fact that ports are, in some ways and by their very nature, cosmopolitan spaces. In other ways, they are also their own worlds and, therefore, contained and closed. What happens here stays here, and so a normally gruesome crime may be seen through a lens that is even more intense and violent.

On 18 March 1984, two armed vigilante groups clashed in Fatehpur Village Road in the Garden Reach area. One S. Singh was the officer-in-charge of the Garden Reach police station, but, on that particular day, the Deputy Commissioner (DC) of Police Vinod Mehta was in the area and led a team to confront the evolving situation. When Mehta arrived at the scene, police were already conducting raids in its narrow serpentine by-lanes. Armed people in the area clashed with police and, when the police fired several rounds, two people were killed on the spot. Mehta and his bodyguard became separated from the larger group and disappeared into the narrow lanes, but how and why they were separated remains a mystery. Unanswered questions also include: Did the constables refuse to follow their superior officer? And were Mehta and his bodyguard lured into a death trap?

Around 1.30 pm, news reports stated that over the last couple of hours Mehta and his bodyguard Mokhtar Ali could not be found. The fact that such a high-ranking officer as the DC (Port) had gone missing in an area where a little while back there was such a fierce clash between the police and a mob of armed miscreants was a matter of great concern. Police

Commissioner Nirupom Som rushed in with a huge battalion of reinforcements from the Lalbazar. Among others Ranu Ghosh, the District Magistrate of the 24-Parganas, was also brought in, and the police began a house-to-house search. However, it was only around 4:30 pm that a tip-off call told police that Mehta had been murdered and his body was stuffed in a drain. The caller gave the location of the body to the police and, when they followed up, they found a devastatingly mutilated body that clearly bore evidence of sustained and brutal torture. Mokhtar Ali's body was discovered in a sack that was set on fire. Such brutal murders shocked the nation. One Idris Ali was arrested as the chief perpetrator, along with four others, two of whom—Nasim alias Naso and Lokeman Shah—were given the most stringent punishment.

For days after, newspapers carried stories as to how the DC (Port) followed by his bodyguard tried to shelter from the marauding hordes. The mob chasing Mehta had been informed that he had tried to defile the mosque. Mehta, in fact, had done no such thing, and when he tried to enter the house of a Mullah for security, the Mullah—in fear of the consequences of giving him shelter—asked him to leave. Mehta and his bodyguard then took to the by-lanes, but the mob spotted them. Mehta entered the house of a police constable and Mokhtar entered someone else's house, but that did not give them any security and soon the mob was upon them with tragic consequences.

Following this, the horror of Mehta's and his bodyguard's death was on the front page of every newspaper for days in the city. Then police went looking for retribution in the Garden Reach area where there were nightly raids and mass arrests. Forty people and nine children were booked. Islamic organizations in the area got together and petitioned the government to stop what they argued were tantamount to blatant human rights abuses. There were also allegations of police molesting women (Sen 1984).

Criticism of the Calcutta police continued for days. Idris Ali, the chief suspect, was found dead in police custody, and there were allegations that he had been beaten to death by the constables. However, that did not alter the criticism that it was cowardice that stopped the rest of the police force in Garden Reach from following Mehta into the by-lanes. Apart from suspending a few lower order policemen, the state government did not take any administrative actions against the police. With the killing of Idris Ali, even the court case against the others lost much of its steam.

The session courts awarded the death penalty to both Naso and Lokeman Shah. In the High Court, however, Shah's penalty was reduced to life imprisonment, but both parties appealed to the Supreme Court where Justice K. T. Thomas, who was on the bench, described the murders as a product of 'communal frenzy.' While in the case of Naso there was an eye witness that saw him dealing a fatal blow to Mehta, the case against Lokeman Shah rested on his confession, and his lawyer, Shri A. K. Ganguli, argued that the confession was coerced. Ganguli cited the case of Idris Ali to show that police were extracting confessions from people that they had already deemed as guilty. In his ruling, Justice Thomas characterized the case as 'the rarest of the rare' and suggested that the rioters had been fanned up by 'literate leaders' who had kept themselves away from the crime scene. Although the judge was sympathetic to the ignorance of the 'demented' rioters and so reduced the death penalty, in consideration of the heinous nature of the crimes, he had to give a sentence of life imprisonment to both Naso and Lokeman Shah (Lokeman Shah and Anr vs State of West Bengal 2001).

From the beginning, however, this case was connected to much larger issues such as the increasing insecurity of minorities. It brought into focus the polarization of the city into new and old migrants; the conflation of caste, class, and religion; and issues of fractured identities. It also exacerbated tensions between the upper and lower echelons of the police. The lower echelons thought that they had to serve the cause of law and order even at the cost of their lives, whereas those at the helm took the kudos. The administration, including the justice system, blamed the events on communal politics without exploring any other socioeconomic reasons for the burning antagonism. There was no discussion of depressed wages, as a result of which there were dock workers' strikes in 1979–80. Until 1975, there had been hardly any strikes, but, after 1975, there was a growing antagonism between the stevedores and the dock labourers, a growth in vigilantism, and the image of police as partisan—all of which contributed to the mistrust between common people and the administration. For decades, this case has remained central to people's psyche whenever crime is discussed in the city, so much so that all port crimes of any magnitude are always compared to it.

THE PORT AND ITS VICINITY

The port of Kolkata is India's longest operating port as well as its sole major riverine port. During the time of the Sutanati traders beginning in the seventeenth century, the port was operative seasonally, that is, between

September and March. The port area was developed after Wajid Ali Shah was exiled there in the early nineteenth century and when Wajid Ali Shah's son revolted against the British, Wajid Ali Shah was imprisoned in the Old Fort William. The British came to Bengal as traders and so it was under them that the port began its ascendance as an area of importance.

The port area is largely populated by immigrants who can be coerced to work as cheap labour and this phenomenon has been taking place for over a hundred years now. According to Jayanta Gupta and Kaustav Roy (2003), a 'lot of people, most of them from Bihar and Uttar Pradesh, settled in the area to work as casual workers. The "sharks" were quick and under their leadership, associations were formed. Where work was unavailable, extortions were resorted to. Labourers were quick to change allegiance from one union to the other, depending upon the situation.'

According to discussions with anonymous respondents in the Garden Reach area, among those living in this area are people from Bihar who travelled to East Pakistan, where they were treated as stateless. Many of them then returned to Kolkata without proper papers from Bangladesh. In the Garden Reach area, there are many who live in abject poverty alongside people who might have been poor once but now are wealthy, mostly because of nefarious activities based on the port. In addition, many of them have allegedly nurtured armed gangs to protect their activities. Consequently, the Calcutta port has long been closely associated with the crimes perpetrated in the city.

A short study of the demographic profile of the region based on data from the Census of India (2011) helps to answer questions such as: What makes this region so vulnerable to crime and criminals? Is there a logistical imperative for this? And why do crimes committed in this region assume such a central place in the city?

According to census data, the population of the port area is overwhelmingly male, and the presence of men far outnumbers women within families and in public spaces. The average household size is 5.6, that is, every household has between five and six people of whom there were many more men than women. It is also, we are told, an overwhelmingly Muslim population. Most small businesses are conducted by men and women work largely behind scenes, although it is also true that the port employs its share of women. As indicated, large numbers of people living in the wards do not have the most basic literacy which makes it difficult for them to get jobs of any quality. Some work in household industries but, for most of them, the docks and port area are their most likely employers.

The census also indicates that the number of non-workers is also much greater than the number of workers in the port area. Therefore, vagrant youth groups are a common phenomenon in the region and vulnerability as a minority population is compounded by large-scale illiteracy and unemployment. There is thus an availability of young men who can be hired as goons. I am not claiming that this is what happened in either 1984 or 2013, but there was a possibility of that happening, particularly because, in the earlier year, there was depression and dock labourers were threatening to strike because of low wages.

Another fact shown by census data is that there is a paucity of schools for children in the port area. A few children might go out of area for schools, but the previous indicators suggest that most of the families do not have the resources to send their children to expensive private schools. Some children could be in household employment but clearly not all. In the Mehta case, nine children were arrested and sent to juvenile homes. None of the above indicators mean much individually, but, when taken together, they indicate why this area might be perceived as crime-prone.

No analysis of port logistics is complete without a discussion of crime. Where else but in ports do we get access to the outside world? Let me now explain why for understanding the nature of activities carried on in the port area it is important to have a sense of the crime scene in this area. The routes through which goods and ideas move are often the same routes taken by organized crime networks and groups. Legitimate business hubs often transform into markets for smuggled goods and, sometimes, the players in both remain the same.

There are very few writings on criminal activities in ports and docks from the perspective of logistics. Many more have been written on ports and labour unrest; however, Clive Emsley (2015) has recently published an article on the criminal activities of British dock labour during the Second World War. Although Emsley connects crime in the docks to the development of the security system in Britain, he does not connect it to logistical networks. My intention, therefore, is to look at the logistics of port crimes and how that transforms the peripheral into the central question of enquiry. A port is often at the periphery of an urban logistical system, but sometimes it can transform itself into the main logistical hub for the development of the urban space. When that happens, crimes in the port become central to the security of that urban space and this is, arguably, true of Kolkata.

History of Port and Crimes

Crimes in the port can take many forms. They can be crimes perpetrated by labourers who pilfer from the goods that the ships bring in or take away, and which they are meant to load or unload. The traditional perception of crime in port areas is largely related to containers, either the cargo they bring in or that which is pilfered from them. Sometimes ship crews and security people are also involved in these thefts. Port crimes can also be about organized gangs making entire containers disappear, thus causing multiple security hazards. In today's world of increasing fear over terrorist attacks, the disappearance of containers can be a serious threat to the state's perception of its security. This problem beleaguers many important ports including in Canada, the US, Australia, Dubai, and India.

What is not so predictable or routine are crimes where the port acts as the backdrop. These can be about issues not directly related to the docks but which are perpetrated in the backyards of the docks or areas adjacent to the port areas. Why port areas are considered to be convenient locales for such crimes needs to be considered. While it is not my intention to list all the crimes that take place in port areas, I will discuss crimes that occur in the port vicinity but have ramifications beyond the port itself. An understanding of such crimes will hopefully also offer some insight into the logistical implications of what causes the port area to become a crime site. Even when crimes appear random, there is always some logic behind why they occurred in a particular logistical space.

Even before Indian independence, criminals from the port area appeared in Intelligence Bureau (IB) files. In the case of Mohammad Musa, he was born in 1920 and came from a poor background. Although he did not receive any formal education, he was very creative and taught himself tailoring. In 1939, at the age of 19, he left for Singapore where he remained until its fall in 1942. Initially, he became a volunteer in the Indian Independence League, and then he went to Penang where he allegedly received training in espionage. He returned to Kolkata in September 1943 and, like many of his neighbours, joined the Port Trust. The IB files report that the Port Trust Workers' Union immediately involved him in spreading anti-British propaganda, especially among the port's Muslim workers. They also indicate that he was involved in 'revolutionary activities,' although it is never spelt out what these activities actually were. Apparently, this young man created so much trouble that the military intelligence considered him to be dangerous and the British government classified him as 'black listed' (IB File No. 473/46).

Another case in the IB files was titled 'Incidents involving the use of firearms' and it happened in mid-1947, a year after the Great Calcutta Killings, which were days of widespread riot and manslaughter between Hindus and Muslims. In this case, it was reported that the Chief Presidency Magistrate had asked the Port police station to conduct raids at 4A George Terrace, a residential quarter, on 28 July 1947. This was mostly likely undertaken because of some connection to the Great Calcutta Killings, and so SI Upendra Chandra Dey took eight constables and conducted the raid. At the time, 4A George Terrace was occupied by Ghulam Mohiuddin who was a Member of the Legislative Assembly (MLA) of Bihar. It is unknown as to why an MLA from Bihar was living in Bengal; however, the search was conducted in the presence of two witnesses—Ramzan Ali and Mohammad Khalid—both of whom were tenants at 4A George Terrace.

After completion of the legal formalities, the police found two guns which were allegedly used in the recent riots in Patna, in Kidderpore, and also in other areas under the Port police station. The guns were seized, including one double-barrelled gun and one single-barrelled rifle, but Mohiuddin told police that the guns had been stolen from him a few months earlier, and he could give no reasonable answer as to how he got the guns back or where he found them. When the police asked him why he did not inform police about the theft, he could not give any plausible answer. The Assistant Commissioner of the Intelligence Branch marked the MLA as 'suspicious,' referred the case to the CID, and advised that he be investigated in Bihar (IB File No. 419/47).

A third case is related to labour problems that beleaguered the port from time to time. A. S. K. Iyenger, General Secretary of the Madras Harbour Workers Union, wrote a letter to J. M. Kaul of Calcutta Port Trust Employees Union in which Iyenger attached a telegram he had sent to Pandit Nehru about the labour strike that was taking place in Kolkata. In the telegram, Iyenger wrote: 'Twenty two thousand Calcutta Port Trust workers went for a strike from 5th of February for a minimum wage. No retrenchment. Example of Hindu-Muslim unity. Please intervene to settle the strike.' In his letter, Iyenger asked Kaul to take the initiative and talk with the leaders of the National Government at Kolkata as well as in Delhi. According to Iyenger, the new national government had to be made to take strong steps to fulfil the demands of the port workers as soon as possible. These demands included (i) minimum wage of INR 40 for workers and INR 80 for clerks, (ii) three months of wages as bonus, and (iii) no retrenchment. The point of retrenchment was a prominent demand because of a retrenchment notice issued by Thomas Elderton, the Chairman of Calcutta Port Trust, on 1 February 1947 (IB File No. 43/47 (I)).

From the IB files, it is clear that the IB followed Kaul's progress over the years. They considered his position as sensitive and also kept tabs on his relationship with other political personalities. In the IB files, there is a letter from Bhupesh Gupta, the Communist Party of India leader, to Kaul, dated 14 January 1957, which pertains to the local elections. In it, Gupta advises Kaul to print election pamphlets in large numbers and simultaneously in three languages—Bengali, Hindi, and Urdu—to reach all the workers of the Port Trust. He says that the pamphlet should also be distributed to the jute mill workers in Garden Reach area and workers of other manufacturing agencies. Gupta asks Kaul to conduct meetings with the workers of the port secretly to try to involve them in election campaigns (IB File No. 43/47 (II)).

Both the British government and the Indian state considered people with radical opinions and political inclinations as problematic and threatening. Musa and Kaul were both such personalities. A further disturbing threat from the point of view of the British or Indian state was the 'outside' factor and Musa had contacts with a world beyond Bengal, as did Mohiuddin and Kaul. Also, a peripheral port crime such as carrying arms in the port area became way more dangerous when the state could establish association between that and the larger politics of the state. This is obvious from the IB reports where perhaps the most criminal of all offences were attempts to radicalize port labour and anyone who attempted that was considered to be extremely suspicious because this impacted the politics of the entire state. For these reasons, therefore, port crimes did not remain localized but were considered to be linked to the larger issues of state politics.

Labour, Theft, and Strike

John Connell writes the following about the labouring class living in another rapidly urbanizing port city: 'The nation's poor have a number of common problems. They live in unhealthy homes with little or no sanitation; they are increasingly exposed to crime; they feel ignored by the state, service institutions and their own communities; their children are exposed to child labour; their children are more likely to be sexually abused by adults; they are deeply cynical about politicians and politics in general' (Connell 2003, 243).

From the late 1960s onwards, the same phenomenon was the plight of dock labourers living in Kidderpore, Metiabruz, Wattgunje, and other areas near the port, but there were certain particularities about the dock labourers. Men and women from the same families often worked together in the docks and many of them could be part of the supply chain of smuggled goods that were later sold in or near Fancy Market, a well-known market for smuggled goods in Kidderpore. There was also a growing hostility towards new immigrants who were seen as bringing wages down as they were willing to work for less. A dock labourer's job was never secure and there was confusion as to who employed them. As late as 1988, there was a writ petition challenging the capacity of the Dock Labour Board to act as the employer of the labourers (Chatterjee 1988). Although the High Court dismissed the writ, it showed the uncertain world of the dock labourer. Therefore, it is hardly any surprise that moonlighting for dock crime syndicates might seem attractive to the labouring poor. Also, people belonging to kith-and-kin networks operated very closely with one other, especially when families were involved in crime.

In his study of crime in the work place, Gerald Mars (1982) divides people committing such crimes into representative groups which are given descriptive animal names. Workers in the dock area who fiddled or performed thefts were called wolves because often they operated in a pack that assured complete confidentiality and that is why it was extremely difficult to apprehend them. A spectacular car theft racket, where it was obvious to everyone that insiders were involved but no one was caught, showed how such smuggling resisted disclosure.

In December 2008, a man was intercepted in Singapore who was allegedly involved in a racket smuggling automobiles into India. After interrogating him, the Singapore officials established that he had recently sent a beige colour Perodua car into India. The man informed that the racket went on through the Kolkata Port, but he did not or could not name any of the operatives in Kolkata other than the buyer who lived in Nepal and someone who was meant to send him the car. On getting the tip-off, the Singapore police informed the Delhi police, who in their turn called the Kolkata police. When the consignment reached Kolkata, the police intercepted it and informed Singapore, but they had to wait two months before a team from Singapore could come and verify that this was the same car.

They were assisted by the Motor Theft Squad of Kolkata to match the number of the engine to ascertain that it was the same car. The Singapore team informed their counterparts in Kolkata that before this incident at least seven cars and some bikes had been sent by the same gang through the Kolkata Port. It could not be ascertained who from the Kolkata side was implicated in this because, by the time investigations started, it was already too late to catch anyone. When I had informal discussions with police officials, who wished to remain anonymous, in January 2016, I was told that they could not make the workers open their mouths even by coercion. When *The Indian Express* (2009) reported on these affairs, they titled the story 'International Car Theft Racket Intercepted in Kolkata.' A year later, *The Times of India* (Bandyopadhyay 2010) revisited the story and presented Kolkata as the centre of a racket whose tentacles were spread across South and South-East Asia. Once again, a port crime was transformed into criminal activity that was not localized but had implications for the entire region, and something that took place in the periphery or the port transformed itself so that the criminal activity of the port assumed a centrality in the city.

Perhaps a much greater crime by the dock workers, from the perspective of the port administration, was their potential ability to strike or verbalize the threat to strike, but, for this offence, they could rarely be punished. Repeated references to J. M. Kaul's activities in the IB files for all those years show that from the early days the state was extremely ambivalent about labour leaders.

In fact, the earliest attempts to regulate labour laws were not meant to give labourers a steady income but to make sure that there were adequate numbers of labourers to service the docks. From time to time, there were nervous declarations that agitated the city about how unprofitable the Calcutta port was becoming and why it might be closed, but that closure never came. For example, in 1979, the average imported cargo landed per ship decreased for all ships (except those from the US and coastal carriers) in comparison with that of the same period the year before. The same was true for export cargo, and, without exception, dock labourers were blamed for this state of affairs. Timir Basu (1979, 1669) explains that the Dock Labour Board (DLB) had failed the workers miserably: 'Some workmen too are not blameless especially the immigrant workers who are anxious to earn more by allowing themselves to be "double booked" … even at the expense of their physical well-being, indeed, this tendency of labour from outside the state is nowhere more prominent than in Calcutta port.' This system of double booking meant that a stevedore did not have to pay two levies to the DLB but merely paid the worker his/her due. Workers on the other hand had to work for 16 hours straight, which created health hazards, but they were so poorly paid that such hours were necessary for workers and their families to ensure mere survival.

As recently as 2012 there was news that traffic handled by major ports fell by 6.33 per cent and again Calcutta Port was cited as the worst off (Balachandran 2012). Port authorities claimed that nearly 50 per cent of the workers were surplus in the Kolkata Port and, also, that greater mechanization meant less pay for the workers. There are times when all workers decided to take a pay cut so that they could pay their out-of-work compatriots. Even then people called the workers extortionists (Jayanta Gupta and Kaustav Roy 2003). That the workers' conditions remained precarious, even after the merger of the DLB and the Kolkata Port Trust, is apparent from their desperate threats to strike, which occurred as recently as 2015 when such action was averted by assurances from the Union Minister of Shipping that the privatization of ports would not be carried out (*The Hindu* 2015).

Striking is not something that individual workers of the port can easily embark on. It means days of preparation and enormous hardships. That the workers take the easier route of pilfering becomes obvious when one looks at the pending court cases. Their options for a living wage are limited and they can either join the smugglers in petty thefts because the larger ones are beyond their capacity or their non-working family members can join vigilante gangs that are usually sponsored by the rich in their communities.

Gendered Crimes in the Port

One Sunday in January 2014, a young woman who worked for a store in a posh shopping mall on Anwar Shah Road took a taxi with a few friends. Her destination was the Howrah Station. Her friends were dropped in the Park Circus area and the taxi took her on to Kidderpore. Although she requested the taxi driver repeatedly to take her to Howrah Station, he refused and asked her to wait for a bus near a food stall. He said that there are many buses passing along this road that would easily take her to her destination. While she was waiting for a bus, five young men arrived in a car and accosted her. Initially they offered her a lift and when she refused and started walking away they physically pulled her into the car. They then took a detour through the serpentine lanes of the port area and stopped the car. They gagged her, stuck a knife to her head, and took turns to repeatedly rape her while she cried and prayed for her life. At some point in the night, they gathered her broken body and dumped her and her belongings including her phone near Babughat. In a near delirious state, she walked to Howrah Station and called her mother who came and took her to Howrah hospital where she was found to be bleeding profusely. At the intervention of the state government, she was taken to the well-known Belle Vue Clinic (*Time of India* 2014).

The next day the police identified one of her tormentors. His name was Mohammad Hamid, alias Raja, and the police were optimistic they would be able to close the case. Meanwhile, a section of the press started agitating, saying that the incident 'puts a question mark on women's safety under the Trinamool congress government that has been rocked by similar crimes in the last two years, the most recent being of a 16-year-old girl being gang raped twice and set on fire by the aides of the rape accused in Madhyamgram' (*Indian Express* 2014).

Between discussions of how members of the State Commission for Women reacted to the incident, there was a full-fledged politicization of the case. 'Under this regime, a horrendous history of gang rape is being created. It only brings out the government's inability to keep law and order under control and provide security to women,' said Leader of Opposition Surjya Kanta Mishra of the Communist Party of India-Marxist Party (*India Today* 2014). Meanwhile, speaking to media persons 'at Siliguri in northern West Bengal, [Chief Minister] Banerjee said: "I have ordered the administration to arrest the culprits immediately and take stringent action. None of the culprits will be spared." She also announced that the state government would pay for the medical expenses of the victim' (*DNA* 2014). Apart from the fact that this case was very like that of the heinous rape of Jyoti Singh in Delhi in December 2012, this was not the first gang rape case in the labyrinthine streets of the port area. There had been a similar gang rape case in 2006; however, none of the newspapers or the media connected the cases. Nor did they question the safety of women in the port area.

Two social scientists, however, have completed such a study and shown that while reports of rape as a crime are increasing across the entire city, according to police statistics, the rate of rape is decreasing in the port area (Dey and Modak 2015). They write that while the molestation of women is on the rise, rape is decreasing. In order to come to an approximation of the truth, I interviewed women from the area who said that they felt fairly secure, but then most of them rarely left their homes unescorted or alone.

Even when they went to school or work, they went in groups. Furthermore, most of the women did not work outside their homes. Their men worked largely in the garment industry, particularly of the ready-to-wear kind, and the women did embroidery work. My conclusion is that what makes this area problematic is not the number but the kind of violent rape and molestation of women who are considered to be 'outsiders' that takes place in this site. Another notable phenomenon reflected in the 2014 case was that instead of rhetorically and spatially containing the crime in the port area discursively, it became central to what was considered the widespread insecurity of women in the region under a particular regime.

Conclusion

Energy plays a key role in influencing the geopolitical strategies of the world. The disruption of energy supplies could be one of the main causes for destabilization of industrialized nations. According to observers 'energy security in South Asia as well as surrounding regions has become a high priority for sustainable economic growth' (Kalim 2016, 209). It is through the ports that supplies of essential items such as crude oil and other goods enter a country. According to one observer, ports 'appear to be a weak barrier against illegal entry and also the weakest link in the logistics chain; hence their vulnerability' (Gunasekaran 2012, 56). It is this magnifying economic importance of ports alongside geopolitical vulnerability that makes port areas sites with potential for crime. My mapping of some of the exceptional and non-exceptional but symptomatic crimes of the Kolkata Port area reflects that at least in terms of crime the port is never peripheral to popular imaginings but rather central to it. Anything happening in the port does not stay in the port but spreads discursively. One reason for this, perhaps, is that the port cannot be contained as its main function is dispersal. Ports are by nature hubs of multiple flows. Also their contacts with the outside world make them a problematic space from the perspective of security because, in the administrative imagination, all that is threatening to stability comes from the outside, whether it be goods or people. Therefore, the role of the state is not only to tightly control such a space but also to determine how to control a space that is allocated for all kinds of movements.

Such space also becomes the hub for the recent immigrants who look upon the state as oppositional, corrupt, and violent. On achieving some material power, these people rely more on their kith-and-kin vigilantes and reject state mechanisms of security. Hence, their most spectacular clashes are with the police. All of this makes the port space peripheral for development but central for violence and, therefore, central for crime. A general state of violence makes it dangerous for anyone who is perceived as an outsider. These figures face the greatest intensity of violence, whether they are women or police officers, especially when they appear as solitary figures and are ruthlessly attacked.

As the port, though located peripherally, is connected to all the arteries, so is crime. Criminal gangs or vigilantes in the port area can be connected to larger gangs or political configurations. Therefore, once a crime is investigated, other connections come to the fore. These connections can be investigated or not, for example, after Vinod Mehta's death, no one really questioned the role of police or politics. Another connection between the port, crime, and politics is money, which provides another avenue by which port crime becomes central to the political imaginings of the people and administration. It is probably the money factor that makes the port area so politically fraught. Therefore, a simple student body election snowballs into a riot, albeit between groups of the same religious denomination, but ultimately it is a police officer who is killed. Is it merely coincidental that between 2012 and 2015, exactly when SI Chowdhury was killed, there was unrest in the port over corporatization? Having learned from the Vinod Mehta murder, the leadership did not try to shield the culprits beyond a certain point. Yet even then the incident was, at least discursively, linked to much larger issues such as Islamic terrorism.

In trying to understand crime, we get a picture of the port that is highly connected to the political economy of the city, if not the region. The port is also a space that is both national and international. Its fluidity and ambiguity make this space, at least in popular perception, more insecure. Thus, security of the space becomes essential because those who control this space also control international trade and international currency. Because ports involve big money, security in port areas has become a commodity that can be traded and, consequently, there is no singular system of security in India's ports.

Because the state is no longer the sole provider of security, and makes use of services delivered by the private security industry, the security of ports becomes plural and complex. According to one observer, 'security pluralism' makes 'the market for crime control ... a highly competitive one, driven by price as much as quality, and in which profit is a more powerful motive than performance' (Zedner 2009, 90). In such a scenario, security is marked more by control than safety and this increases the possibilities for violence which, when it happens, is transformed into a spectacle to reinforce control. This possibility of spectacular violence makes ports central to the imaginings of crime in the city. Insecurity can result from multiple sources including excessive security and, therefore, the question remains: Is securitization the answer to insecurity? Yarin Eski (2011, 248) argues that 'insecurities should not be fought with more security and securitization, but dealt with by individual and societal resilience.' Whether such a perspective can be considered as a corrective to insecurities of people in the Kolkata Port and its vicinity is a moot point as, even today, the Kolkata Port remains central to big crime in the popular imagination of the city.

REFERENCES

Aajkal. 2016. 'Kahlon: Another 15 Lakhs Recovered,' 11 March 2016.
Balachandran, Manu. 2012. 'Traffic Handled at Major Ports Falls by 6.33%.' *The Economic Times*, 15 May 2012. Accessed 23 February 2016. http://articles.economictimes.indiatimes.com/2012-05-15/news/31711530_1_major-ports-minor-ports-indian-ports-association
Bandyopadhyay, Krishnendu. 2010. 'Kolkata the Hub Of Global Car Racket.' *Times of India*, 1 November 2010. Accessed 22 February 2016. http://timesofindia.indiatimes.com/city/kolkata/Kolkata-hub-of-global-car-racket/articleshow/6849219.cms
Bartaman. 2016. 'Another 15 Lakh was Found with the Port Chairman and the Businessman,' 11 March 2016.
Basu, Timir. 1979. 'Dock Workers of Calcutta.' *Economic and Political Weekly* 6 (October): 1669.
Bhabani, Shoudhriti. 2013. 'Policeman Shot Dead and Four Students Injured After Rival Unions Clash at Kolkata College.' *Mail Online India*, 13 February 2013. Accessed 1 February 2016. http://www.dailymail.co.uk/indiahome/indianews/article-2277692/Policeman-shot-dead-students-injured-rival-unions-clash-Kolkata-college.html#ixzz43WdI2hjS
Chatterjee, Judge Susanta. 1988. 'The Master Stevedores' … vs Calcutta Dock Labour Board And … on 11 October, 1988.' The Calcutta High Court. http://indiankanoon.org/doc/896233/
Chaudhuri, Monalisa. 2016a. 'ED on Kahlon Cash Trail.' *The Telegraph* (Calcutta) *Metro*, 15 March 2016.
———. 2016b. 'Kahlon's Aussie Link.' *The Telegraph* (Calcutta) *Metro*, 16 March 2016.
Connell, John. 2003. 'Regulation of Space in the Contemporary Postcolonial Pacific City: Port Moresby and Suva.' *Asia Pacific Viewpoint* 44, no. 3: 243–57.

Covert Wires. 2013. 'Kolkata Police is Murdered by Islamic Goons in Daylight: Police and Politics Mute Spectator,' 16 February 2013. Accessed 3 January 2016. https://asansolnews.wordpress.com/2013/02/16/kolkata-police-si-killed-by-muslim-goons-in-braod-daylight-impotent-police-administration-and-politicians-mute-spectators/
DNA. 2014. 'Mall Employee Gangraped in Kolkata, One Arrested,' 20 January 2014. Accessed 23 November 2015. http://www.dnaindia.com/india/report-mall-employee-gang-raped-in-kolkata-one-arrested-1954372
Dey, Falguni and Swagata Modak. 2015. 'Crime Against Women in Kolkata: A Spatial Difference and Temporal Change Analysis.' *International Journal of Science, Environment and Technology* 4, no. 4: 1139–52.
Emsley, Clive. 2015. 'Cops and Dockers.' *History Today,* 65, no. 6 (August): 19–25.
Eski, Yarin. 2011. 'Port of Call: Towards a Criminology of Port Security.' *Criminology and Criminal Justice,* 11, no. 5: 415–31.
Express News Service. 2013. 'Garden Reach Violence: Accused held,' 8 March 2013. Accessed 10 February 2016. http://archive.indianexpress.com/news/garden-reach-violence-accused-held/1084929/
Gunasekaran, Periasamy. 2012. 'Malaysian Port Security: Issues and Challenges.' *Australian Journal of Maritime and Ocean Affairs* 4, no. 2: 56–68.
Gupta, Jayanta, and Kaustav Roy. 2003. 'Labour Trouble Gives Kolkata Port a Bad Name,' *TNN,* 6 August 2003. Accessed 10 December 2015. http://timesofindia.indiatimes.com/city/kolkata/Labour-trouble-gives-Kolkata-port-a-bad-name/articleshow/117088.cms
IB File No. 419/47, sub: Incidents involving the use of fire arms and bombs during the communal disturbances in Calcutta. Available in State Archives, Shakespeare Sarani.
IB File No. 43/47 (I), sub: Jolly Mohan Kaul, s/o S. M. Kaul of 23/3 Roy Street Calcutta (Secretary, Calcutta Port Trust Employees Association). Available in State Archives, Shakespeare Sarani.
IB File No. 43/47 (II), sub: Interception of a letter from Bhupesh Gupta, C.P.I. to J. M. Kaul, (Memo No. 1042 (2) of 14.1.57 at Park St. P.O.). Available in State Archives, Shakespeare Sarani.
IB File No. 473/46, sub: Mohd. Musa s/o Mohd Daud of Metiabruz, 24Pgs. Available in State Archives in Shakespeare Sarani, Kolkata.
India Today. 2014. 'Mall Employee Dragged Out of Cab, Gangraped in Kolkata; One Arrested,' 20 January 2014. Accessed 23 November 2015. http://indiatoday.intoday.in/story/woman-dragged-out-of-cab-gangraped-in-kolkata/1/338802.html
Indian Express. 2014. 'Kolkata: Shopping Mall Victim Gangraped,' 20 January 2014. Accessed 23 November 2015. http://indianexpress.com/article/cities/kolkata/woman-gangraped-inside-truck-in-kolkata/#sthash.m8SGa8RZ.dpuf

Kalim, Inayat. 2016. 'Gwadar Port Serving Strategic Interest of Pakistan.' *A Research Journal of South Asian Studies* 31, no. 1: 207–21.
Lokeman Shah and Anr vs State of West Bengal on 11 April, 2001, in Bench: K.T. Thomas, R.P. Sethi, S.N. Phukan, Petitioner, Supreme Court of India, CASE NO: Appeal (crl.) 784 of 2000; Appeal (crl.) 785 of 2000.
Mars, Gerald. 1982. *Cheats at Work: The Anthropology of Workplace Crime.* Winchester: Allen & Unwin.
Roach, Ashley. 2003. 'Containers and Port Security: A Bilateral Perspective.' *International Journal of Marine and Coastal Law* 18, no. 3: 341–61.
Sen, Saibal. 2013. 'Angry Cops Drafted FIR to Deny Iqbal Escape Hatch,' *TNN* 19 February 2013. Accessed 2 February 2016. http://timesofindia.indiatimes.com/city/kolkata/Angry-cops-drafted-FIR-to-deny-Iqbal-escape-hatch/articleshow/18566692
Sen, Sumanta. 1984. 'Police Paralysis, Vinod Kumar Mehta Murder, Idris Mian Custody Death Show Calcutta Police in Poor Light.' *India Today,* 15 May 1984. Accessed 14 November 2015. http://indiatoday.intoday.in/story/vinod-kumar-mehta-murder-idris-mian-custody-death-show-calcutta-police-in-poor-light/1/361303.html
The Hindu. 2015. 'Port Union's strike called off,' 15 March 2015. Accessed 22 February 2016. http://www.thehindu.com/news/cities/Kochi/port-unions-strike-called-off/article6995367.ece
The Indian Express. 2009. 'International Car Theft Racket Intercepted in Kolkata,' 1 April 2009. http://archive.indianexpress.com/news/international-car-theft-racket-intercepted-in-kolkata/441631/. Accessed 15 January 2016.
The Telegraph. 2013. 'Munna Named Murder Mastermind,' 4 May 2013. Accessed 15 January 2016. http://www.telegraphindia.com/1130504/jsp/calcutta/story_16855333.jsp
Times Now. 2013a. 'Law Will Take its Own Course: Mamata on Garden Reach Violence,' 15 February 2013. Accessed 2 February 2016. http://www.timesnow.tv/Law-will-take-its-own-course--Mamata-on-Garden-Reach-violence/articleshow/4421136.cms
Times Now. 2013b. 'WB Guv critical of minister's alleged bid to shield accused,' 15 February 2013. Accessed 2 February 2016. http://www.timesnow.tv/Law-will-take-its-own-course--Mamata-on-Garden-Reach-violence/articleshow/4421136.cms
Times of India. 2014. '21-Year-Old Girl Gang-Raped for 2 Hours in Kolkata, 1 Held,' 21 January. Accessed 2 February 2016. https://timesofindia.indiatimes.com/city/kolkata/21-year-old-girl-gang-raped-for-2-hours-in-Kolkata-1-held/articleshow/29127062.cms
West Bengal Series-20, Part XII, District Census Handbook, Kolkata Village and Town Wise Primary Census Abstract (PCA), Census of India, Directorate of Census Operations, West Bengal, http://www.censusindia.gov.in/2011census/dchb/1916_PART_B_DCHB_KOLKATA.pdf
Zedner, L. 2009. *Security: Key Ideas in Criminology.* Oxford, Routledge.

CHAPTER 5

Haldia: Logistics and Its Other(s)

Samata Biswas

Contrary to the commonsense that understands logistical networks in terms of products, money, and impersonal chains and flows of information, recent research has focused on the human side of logistics (Neilson and Rossiter 2014). This chapter investigates the interaction of existing 'big' logistical systems with human action, and details alternative logistical networks that mirror big logistics, but also point to the gendered profile of paid logistical work. From the margins of the Bengali port city of Haldia, I focus not on the capital associated with big logistics but on how its excesses and spillages create networks of their own. Women, children, and the elderly populate this alternative logistical world, but—unlike the machinery, money, and movements that characterize male labour in Haldia—their work gets swept under the carpet in the predominant logistical vision of the port city.

HALDIA: HISTORY AND THE PRESENT

Haldia, on the banks of the rivers Hooghly and Haldi, in southern West Bengal (India), was chosen in the 1950s as the site to build an extension of Calcutta port,[1] owing to the various problems plaguing this ageing

---

S. Biswas (✉)
Department of English, Bethune College, Kolkata, West Bengal, India

© The Author(s) 2018
B. Neilson et al. (eds.), *Logistical Asia*,
https://doi.org/10.1007/978-981-10-8333-4_5

piece of infrastructure. An article titled 'To Reharbour Calcutta' (Nair 1971) outlines the crisis that Calcutta port was undergoing and points at measures, both proposed and adopted, to rectify the situation. The problems outlined were of two kinds. The structural difficulties of the Calcutta port included a loss of draught due to silting, while the functional difficulties included the under-utilization of existing dredging facilities. The port had become a pilferer's paradise, the level of efficiency was dubious, and there were immense berthing delays. There were also far too many employees, and this placed a gigantic financial burden on the port (1327–28).

Apart from requisitioning the Farakka Barrage (to be built on the river Ganges, of which the Hooghly is a tributary), with the express purpose of channelling large amounts of water into the Hooghly to flush out the sedimentation at Calcutta port—and continual extensive dredging (with the Government of India, the Calcutta Port Trust, and the users sharing the cost)—the construction of Haldia Dock Complex (HDC) was the envisioned solution to the structural difficulties at the Calcutta port. HDC began its operations in 1977 and, from the time when it was first envisioned in the 1950s to the first assessment studies of 1959–60, Haldia's locational advantages were cited as the main reason for its suitability: there are only two sandbars en route to Haldia and the draught could allow vessels longer than 530 feet, which was the longest allowed in Calcutta. Haldia was also free from the bore tides that afflict Calcutta. It was on the western bank of the river, which provided easy railway and roadway connections to portions of the port's hinterland that were the major centres for coal and ore mining, including iron and manganese.

Indian ores had lucrative export markets and the industry required the facilities of a modern port for the large ore carriers. Similarly, increasing oil imports necessitated a modern and deep draught oil jetty for oil carriers. The Haldia Development Authority (HDA 2016) website and *Haldiar Itikathaa* (Brahmacari 1986, 2) quote an expert, R. O. Gross, who argued that more than 75 per cent of total seaborne cargoes across the world consisted of oil or dry bulk commodities that were carried in ships of increased size. Hence, planning was necessary for construction of a new dock system at Haldia, midway between the sea and Calcutta Port.

The mere building of the dock complex was not enough. The entire region was slated to become a 'hub' of industrialization, with factories that would directly benefit from the newly constructed dock. Roads and railways would carry goods and raw material to and from the dock to the vast hinterland, and raw material and finished products would travel to and from the factories. All of this required massive tracts of land and, over a decade, the Calcutta Port Trust acquired huge areas on the banks of the Hooghly and also reclaimed large portions of land from submerged areas. The Haldia Municipality website demonstrates a linear vision of progress and development, through industrialization, that was crucial to the formation of modern-day Haldia. It charts the movement from agriculture to industry, from dirt roads to concrete ones, and from health centres to hospitals:

> Haldia Municipality took its birth on 9th June, 1997. From a small village, it has today turned up into an industrial town. From merely agricultural land, it has transformed into a land of livelihood based on industrial economy. The developmental manifesto of the Municipality is also changing day by day based on social and economic upgradation. From clayey road to Morum, from Morum to Bitumen, from Morum to concrete roads – changes after changes. From hurricane to electricity, from dry well and a few tube wells to water pipe line connection at household level, from a mere village health centre to sub-divisional hospital and decorated nursing homes, modern dental institute with sophisticated equipment and specialized physicians. (Haldia Municipality 2016)

The 761 square kilometres of Haldia Planning Area have widely varying agriculture, occupations, human habitation, pollution levels, and vegetation. Haldia is at the same time a city, a municipality, a riverine port, and an industrial belt—each standing in for the other and, according to the municipality website, each working in complete cooperation with the other to realize its stated vision. This cooperation or, if one were to term it such, collusion can be seen from the structure of both the HDA and the Haldia Integrated Development Agency Limited (HIDAL). The HDA was set up in 1979, the website claims, for the task of 'turning Haldia into a fast track high growth centre for industry' (HDA 2016). Apart from the local Member of Parliament (MP), the local Member of the Legislative Assembly, and various bureaucrats, the HDA counts as its board members the Executive Director of Indian Oil Corporation Limited (IOC), Haldia, the Managing Director of Haldia Petrochemicals Limited (HPL), and the Deputy Chairman of KoPT, HDC. The HIDAL was incorporated in 2015, under the Companies Act of 1965, to implement projects sanctioned by government departments and ministries and to develop and implement infrastructure projects in the Haldia region. Its key sponsors are also local industry giants such as Infrastructure Leasing and Financial Services, Exide, Dhunseri, and HPL.

Any attempt to map the logistical networks that criss-cross Haldia—and quite literally so, in the forms of warehouses, pipelines, roads, trucks, local transport, container traffic, human beings, and cattle—has to pay significant attention to the interconnections between industry, the dock complex, the geographical area, human actors, and their nonhuman interests.

In its present state, HDC has three oil jetties, four mechanized terminals, six multipurpose terminals, and two container terminals. Other facilities are a port railway, with its own signalling system and port-owned locomotives, and a siding area for general and bulk cargo. For liquid cargo, HDC offers storage facilities, both inside and outside the custom-bonded area, equipment support, power usage, and computer support (as is the case with other kinds of cargo). Other advantages advertised by HDC include rail connectivity and excellent surface transport. The potential for downstream industries in Haldia is also great, especially since the construction of the Haldia Petrochemical plant and arrival of other multinational corporations like Mitsubishi. Also proximate are various other players such as Indian Oil Corporation and Oil and Natural Gas Corporation Limited.

On paper, all of this suggests that Haldia is a place where industry and the port system have absolute synergy. However, there are some structural inadequacies, because, despite dredging, the draught at the port never reaches the promised depth. As a riverine port, ship movement in Haldia is dependent on the high tide as the lock gates can only be opened at specific frequencies. This, and the presence of only a few multipurpose berths, makes the berth waits long and ship scheduling becomes a challenging problem (Kumar 2011). Even though ships are allotted berths on a first-come, first-served basis, unless other exigencies occur, the wait for berths has been reported as unnecessarily long. The death knell for Haldia was rung as early as in 1976 when a commentator in the *Economic and Political Weekly* (1976, 1386) observed 'Like Kolkata port which is dying, Haldia too will die.'

From conversations with different stakeholders at HDC and through an extensive survey of relevant newspaper reports, three key areas of concern emerge: draught and dredging, labour unrest, and charges of political manipulation. Everyone concerned with the running of the dock stresses the importance of maintenance and dredging to keep the existing Auckland Channel functional throughout the year. Dredging to increase the draught of the existing channels seems unlikely to occur as such dredging is subsidized by the Central Government. A former HDC employee (now with

Five Star Logistics) has alleged that the withholding of funds for these purposes will encourage ships to use Dhamra port—a private port in Orissa (built by Tata Steel but later bought by the Adani group from which the current Bharatiya Janata Party (BJP) government gets cut backs). It has also been alleged by locals that, in order to bring down subsidies, the newly discovered Eden channel is being favoured even though it has a draught advantage of only 0.2 metres over Auckland. The Eden Channel can be used more during the monsoon when pilotage becomes more difficult due to the constant motion of the sea. Among other means adopted to bypass the draught situation, there is a sizable transloading facility at the Konica Sandheads (approximately 70 kilometres from the confluence of the Hooghly) where cargo is unloaded and then brought to HDC on barges or smaller vessels. There are also plans to build a deep draught port facility at Sagar Island, with 74 per cent of the expenditure to be met by KoPT and the remaining 26 per cent by the state government.

The General Manager of HDC, Amal Dutta, was keen to direct my attention to these initiatives as well as various other innovations, infra-structural developments, and outreach programmes undertaken by HDC. His anxiety to ensure what he deemed to be a fair and impartial understanding and consequent representation of HDC and its activities was apparent in his haste to brush aside all negative comments presented in the media. A concerted analysis of media mentions of Haldia between April 2015 and April 2016, however, did not testify to Mr Dutta's claims of a resurgent HDC. Instead, most of the reports had to do with labour unrest, corruption, and industrial shut downs. Unlike the Kolkata port, HDC does not draw its labour from the Dock Labour Board. Instead, along with its shift towards a partial landlord model (Hill and Scrase 2012), some berths are leased to private players and most of the workers at HDC are contract employees who have often been subcontracted.

After 34 years of rule by the Left Front, led by the Communist Party of India (Marxist) or CPI(M), West Bengal (including Haldia) has witnessed a change of guard, with the coming to power of the Trinamool Congress (TMC). Within a few years, many trade union members, who had been the unquestioned bastion of the Left, have changed allegiance—a move that was mirrored in Haldia. Leaders of the currently powerful dock employees suggest that post the change of power in 2011, HDC has had all its problems solved amicably and peacefully. In reality, we are faced with the almost unbelievable story of the exit of the private cargo handler, Haldia Bulk Terminals (HBT), from HDC. HBT claimed that as a mechanized cargo-handling operation, it was not being given the cargo it had been promised. KoPT, by contrast, denied making any such promises, and following protracted negotiations, HBT quit Haldia while citing abduction and threats to the lives of its employees. The union leaders responded by claiming that HBT wanted a way out, after quoting absurd amounts in their tender bid, and their exit from Haldia was just way of saving face.

The 'duping' of KoPT by Ripley & Company is another relevant news story from the recent past. A logistical and stevedoring company, Ripley was accused of cheating the West Bengal government of an estimated INR 245 billion during the 34 years of Left Front rule (*The Pioneer* 2014). As a result, the company was blacklisted at HDC. According to a Public Interest Litigation filed at Calcutta High Court, Ripley—despite realizing an estimated INR 3–5 billion per year—had not paid any royalties to Haldia in 34 years and paid only about INR 5000 as a licencing fee for the entire duration. These developments led to the imposition of royalties for cargo handled at Haldia (INR 14.77 per tonne) and future implementation of this scheme across the country. However, in less than a year, Ripley found its way back into Haldia port in partnership with Orissa Stevedores Limited (OSL) and Bothra Group. Despite being major players in the ports of Vizag and Paradip, both OSL and Bothra chose to work with Ripley, calling it a 'natural partner' and claimed it was in the best position to 'understand the ground reality in Haldia' (Saha and Phadikar 2016). The willingness to share profits renders this arrangement suspicious. Port users fear that, given that all key players have become each other's partners, the competitive pricing system that was meant to be ushered in with the introduction of a tender process will no longer be possible. Some users have also pointed to the direct involvement of the current MP in brokering the deal. News reports (see, for instance, Gupta and Mondall 2016) indicate that this same MP claimed that the union was not interested in interfering with matters of shore handlers.

NOTES ON FIELDWORK: CENTRES, MARGINS, AND DISPLACEMENTS

In relation to method, my approach to studying logistics in Haldia has been twofold. As someone living and working at Haldia, I could access key figures in the logistical networks of Haldia (such as labour leaders, contractors, and port officials) and thus claim knowledge of the 'official' discourses surrounding logistics in Haldia. I could also interact extensively with marginal populations—such as logistical workers, land-losers, farmers, and shopkeepers—to access the other side of the official narrative and inquire into the human aspects of logistics. In doing so, I inevitably faced the question of gender—both my own as a researcher and that of the people I was interacting with. I also came to recognize the difference between the 'big' logistics that structures the space of Haldia and its 'others'—including marginal women, children, and the elderly. Fieldwork was carried out, between April 2015 and April 2016, involving ethnography, and the study of newspaper articles and popular magazines published in Haldia, documentation of logistical operations, and accessing documents related to HDC.

Gender was uppermost in my mind while interviewing Shaikh Mujaffar, Chairman of Five Star Group (an umbrella term for several companies based in Haldia that are connected to port and shore handling of cargo), who I met at the Shramik Bhavan (formerly the headquarters of CPI(M), which was vandalized after the TMC win and is slowly returning to use as a place for left trade unions to gather under Mujaffar's protection). Apart from the marginal women I mention in the third section of this chapter, almost everyone I spoke to in the course of this research was a man. I met with these interviewees at the dock, in an HDA flat, at the office of the newspaper *Apanjon*, in the HDC office, and at the Five Star warehouses. All were spaces in which women were absent. This absence of women, in itself, should not be a surprise, as traditionally ports have been male spaces. As Nelli Kambouri (2014) explains: 'It may seem that dock work is stereotypically normalized as masculine only because in the past it required strong hands, but most of all it is the ability to work without having family or domestic care responsibilities that determine the gendered division of labour in the Port' (17). For people whose lives are centred, whether directly or indirectly, upon the dock complex, it is therefore no surprise that spaces would also be strictly coded by gender. In my analysis, I view the space of Haldia as a space made and characterized by male labour, big machinery, and unusual landscapes, but also a space that constantly excludes the labour of women in the making of logistical spaces.

The port city has been studied as a city marked by global flows, and one that is trying to retain its cultural heritage in the face of neoliberal policies and containerization which threaten to disrupt the very fabric of its dock-side existence. The literature also depicts such cities as straddling a liminal space between negative environmental impacts, business growth, and economic development. Managerial perspectives suggest that a variety of stakeholders are crucial to the dynamic and complex innovation networks that characterize port cities and the areas surrounding them (De Martino et al. 2014, 435).

HDC was reclaimed from the erstwhile Doro Parganas where people's primary occupations were fishing and agriculture. While in the beginning it was just an oil jetty, commissioned alongside the Haldia Urban Industrial Complex in 1959, the bare bones of its history do not adequately account for what happened before, during, and after the dock was built. People died in the making of the dock complex and they are commemorated in a park bearing the name of the then MP, Satish Samanta. My ethnographic research has pointed in two different directions: one where people benefitting from the building of the dock went on to earn huge sums of money; and the other in which local populations were displaced, handed a pittance as a rehabilitation package, and ended up with no land to settle on and no job. Haldia Udbastu Kalyan Samiti (an organization run by people displaced by the HDC who use the word *udbastu*, meaning refugee, to describe themselves) has held sit-in demonstrations in front of the HDC office, in a building known as Jawahar Towers, every day since 2014. Needless to say, this dichotomy of development and its unfulfilled promises does not fully capture the complexities of this space, or, for that matter, its originary narratives. If teleology is suspect, so are origins, since both align an analysis ideologically.

In a moment of unexpected poeticism, Tamalika Panda Seth, the former MLA of Mahishadal and Chairperson of Haldia Municipality (the only non-marginal woman I spoke to in the course of my research), recalled the time when if the Port Trust wanted to acquire land, it did so. Was this a harking back to the time when, led by her husband and local MP Laxman Seth, the Left Front government sought to acquire land for a Special Economic Zone at Nandigram, across the river, but was trounced by organized popular opposition? Yes, and no. While ascribing motives to an ethnographic account is the established anthropological practice, the need to read ethnographic accounts as analyzable narratives is also well established now. Tamalika Panda Seth spoke eloquently of a man called Subal Das, a technical hand at the port who was reputed to take great pleasure in pulling the roof off houses with his crane while those living inside struggled to collect their belongings and ran for cover. She never met the man herself, but, for children like her, living in the exact same space where HDC now stands, Subal Das was a feared creature who mothers told stories about when their children refused to sleep or eat. It is crucial to keep in mind that this achievement of mythical proportions was aided by the mastery of new technology, the crane. The crane, with its gigantic size and ability to destroy, must have been an active component in the children's imaginations.

Containers, cranes, and heavy machinery dominate the logistical and visual landscape of Haldia. Their imposing presence, coupled with the understanding that they yield great power, and can be moved and made to comply only with great mastery and technique, makes it a male domain. Crane workers comprise a sizable constituency among the Class III technicians in the dock, the others being mainly electricians, mechanics, and welders. Different kinds of cranes, the Carl Mar, the fork lift, the loader, and the ominously named Hydra, are integral to the logistical worlds I am mapping here—lifting, shifting, dumping, loading and unloading, levelling, laying, and, as has been mentioned already, dismantling.

According to Tamalika Panda Seth, another figure who haunted dismantled homes was *grhalakshmi*, the goddess of these homes, whose sighs and howls were audible from a distance and who haunted the abandoned houses that would soon lay underneath the dock in the Doro Pargana. *Doro*, from *dariyā*, indicates that the silted, fertile land on which the dock was constructed was once under water. In the abandoned empty homes, *grhalakshmi* could be heard in the evenings, crying for obvious reasons. From the aggressive males who displaced *grhalakshmi*, to the future that would come to be dominated by similarly masculine figures, the logistical space of Haldia is marked, although not overdetermined, by the dock complex.

Sentimental narratives do not fully explain the manner in which land was acquired by the Port Trust in the 1960s and 1970s. Tamalika Panda Seth and other respondents accused the HDC of malpractice and of dealing with urban space in a feudal manner. HDC and the industrial belt (which contains around 400 industrial units, having attracted 36 per cent of total investment in Bengal) are inextricably linked with each other in a logistical web, but it is a web that stretches back to the inception of the port. The Port Trust acquired and continued to hold large tracts of land, later selling parts of it to various industries including the two largest: the Indian Oil Corporation and Hindustan Fertilizer Corporation (now closed). Over the decades, it has often been accused by industry players of not subsidizing the licence fee to attract new business and, at other times, there have been lawsuits. It is also engaged in a continuous tussle with local bodies, for example, over failing to purify water for long periods and even, in a stroke of negligence, obtaining tankers of water from elsewhere for the inhabitants of the township (where employees of the Port Trust, IOC, and HPL, among others live) while leaving the rest of the city to fend for itself.

As mentioned earlier, for some, the HDC is what has made them who they are. Mujaffar, for example, shares eloquent reminiscences of the past. He claims to have begun working as a *mati kata* labourer who dug and shifted clay when the dock was being built. He and his brothers, having lost their home and all their land to the dock, were employed by HDC. From 1977, he became close to Laxman Seth, initially a part of the Indian National Trade Union Congress, which lost steam from 1978 onwards when many unions shifted to the Centre of Indian Trade Unions. He became an employee of CPT in 1978 and remained there until becoming a councillor of Haldia Municipality in 2002.

Five Star is now one of the biggest shipping and logistics service providers at Haldia, undertaking loading and unloading work as well as supplying labour, containers, trucks, cranes, and storage space. Anupam, an employee at Five Star, explained to me what logistics meant. According to him, it means handling cargo once it comes off the ship, or from somewhere else, and preparing and loading it for its next destination. While human agency determines when and how the many units of cargo are moved, the most precarious element in this logistical network remains labour. As mentioned earlier, the logistical worker is almost always male: driving trucks, supervising, manually unloading cargo from containers or trucks, shovelling them into sacks, loading the sacks onto trucks again, guarding warehouses, or 'manning' weighbridges. In this logistical world, women are marginal entrants, engaged to clean the warehouses or as itinerant contractual labour to fix small things, mix cement, or lay bricks. In logistical enterprises, apart from a handful of employees with supervisory rank or expert technical knowledge, the workers are contractually employed according to the amount and nature of cargo to be handled. As in the dock, employment at a logistical outfit is largely based on the requirement of workers to be flexible with respect to time and mobility, traditionally considered male attributes.

People who work at such enterprises are local and include some whose lands have been acquired to build giant storage sheds. There are no government data available on the numbers or caste/tribe composition of the sharecroppers and farmers who lost land or livelihoods as a result of the Haldia project (Guha 2008). Nonetheless, with the backing of the now powerful TMC, the agitation by people displaced by the dock is gathering momentum. In a 2005 interview, the then Deputy Chairman of the HDC,

M. L. Meena, claimed 'we have given jobs to one person per family who had lost their homes and land due to the dock complex. Displaced families were rehabilitated further inland' (quoted in Hill and Scrase 2012, 45). Ten years down the line, 40 odd people are still gathered near the gate of Jawahar Towers, living in a makeshift hut and eating from a communal kitchen, in a round-the-clock protest on behalf of 800 families. They are unequivocal in their condemnation of the party that was previously in power. A few of those agitating had been given space as a result of protracted negotiations between HDA and the port authorities, both of which had decided to designate equal amounts of land to set up various residential colonies. But rehabilitation into colonies such as Kshudiram or Gandhinagar does not make up for the agrarian land that families have lost. These colonies, which are now part of the spiralling urban fabric, can accommodate dwellings only, not the land that they worked upon. While the resettlement of those displaced by the port continues to remain an ongoing grievance, the number of shanties across National Highway 41 (NH41) continues to increase. People come, build shanties and small enclosures where they then grow a few vegetables and livestock, build a mosque and a temple, and try to cultivate the surrounding fallow land. Their children then start going to the government-aided schools nearby, and the men find jobs repairing vehicles or in stealing or unloading cargo when wagons stop at the signals. The populations at other established colonies, such as SaotalChak, are also increasing every day.

Port-induced displacement, although large-scale in nature, has not generated significant academic interest in India. Douglas Hill and Timothy J. Scrase hold Parasuraman to have authored the definitive study of displacement with respect to the Jawaharlal Nehru Port at Mumbai, demonstrating wide discrepancy between claimed rehabilitation and actual dispossession. In their 2012 study of HDC-induced displacement, Hill and Scrase claim:

> While there have been financial gain and business opportunities for the lucky few, within the Haldia area there has been widespread displacement, dislocation, job losses and a range of other social and environmental impacts felt by the marginal, local population of peasant landowners and contract, migrant (from other districts) labourers. (2012, 42)

My ethnographic research, coupled with interaction with agitating dispossessed land owners, supports Hill and Scrase's findings. However, Hill and Scrase end their discussion by pointing to the influx of relatively well-off people who are buying land from inhabitants who were originally rehabilitated, and this has contributed to the enormous price jump that current-day Haldia has witnessed. They do not comment on the new influx of economically vulnerable people into Haldia, and their research stops short of examining the existing forms of political involvement in issues of settlement/resettlement despite the change of government.

During the heyday of CPI(M), ICARE,[2] the company run by Laxman Seth, is reported to have bought 37 acres of land from HDA for only INR 37. Haldia Government College's land houses a flourishing colony of people who have 'encroached' into the space, but as the ICARE example illustrates, there is a fine line between encroachment and legal occupation. Migration into Haldia is at least twofold. People come from all over India to work in corporations and, in the case of Mitsubishi, from the rest of the world. These corporate employees are housed in townships or other upscale residential neighbourhoods. They are not considered displaced, although many of them claim that the local population holds them in suspicion, calling them *phnaker manush,* people who have entered through a gap. The rest are itinerant male labourers, who live in rented accommodation and drive the trucks that transport containers to and from the dock, and carry the chief cargoes of coal, iron ore, scrap, and petroleum. Both kinds of migrants are actors in big logistics.

The 'Other' Logistics

Having so far inquired into the development of Haldia as a logistical space with a well-defined hinterland and established industry, it is time to look at the 'other' of logistics: an informal network also dependent upon HDC. I have in mind a population of marginalized, underprivileged, and displaced women who, despite their impoverishment, identify as economic actors within this space of increasingly precarious and short-lived employment. This section of the chapter also explicates how this other of logistics phenomenologically constitutes the predominant logistical vision of Haldia. I investigate spillages, excesses of the cargo handled at HDC, and their relationship, actively aided by human intervention, with the existing logistical framework of Haldia.

Through the figure of Jaya, a migrant woman who makes and sells coal pellets, I ask questions about the difference between male and female labour and its implications for the kinds of spaces female workers occupy, the technologies under their control, and the economic viability of their action. Other, different kinds of subsistence economy can also be witnessed in Haldia. A consideration of the people involved in Haldia's flourishing scrap business could be a case in point. The filling up of port and municipality land with residue generated from a thermal power plant has led to another form of subsistence economy—the building of garages and auto workshops that allow men to eke out a living. Of these, I focus on the case of the female coal dust collectors.

The literature suggests a systematic undervaluation of women's work manifest through occupational segregation, discrimination, and women's unequal share in family responsibilities (Grimshaw and Rubery 2007). In the rural context, a similar observation holds true for the 'invisible' and 'undervalued' contribution of women in agriculture, both as cultivators and as agricultural labourers (Deere 2005). Coupled to this is the devaluation of women's work, the undermining of their contribution to economic production, and the policy lacunae in dealing with women's identity as economic agents, for example, including their identity as 'farmers' (Brandth 1994). The use of the term labourer here is deliberate, to differentiate between and also to draw attention to the unequal distribution of resources and ownership among men and women, which remains disproportionate to the time and effort women put in as part of the workforce. Here, I liberally draw on the information and insights generated by researchers working on valuation of women's work to explain the new forms of employment in which underprivileged and displaced women in Haldia participate. I also explore the poorly paid, never-ending, and flexible nature of this work, and the ways in which these women construct their subjectivities as wage earners and, at times, family wage earners.

Like 22.34 per cent of Haldia's population (Haldia City Census 2011), Jaya is a slum dweller. She lives in the shanties that have been constructed on land that belongs to Haldia Government College. These shanties are relatively new. They were built in the last few years of the CPI(M) regime and have been extended in the past five years. Jaya's shanty is more derelict than many of the others, and the reason for this is probably due to two facts. First, although she is a migrant like the rest of the people in the shanties (or colony, as they call them), Jaya's displacement is far greater than many others, thereby denying her any form of social capital. While some claim to have come from the other side of the highway, from Nandigram across the river, or from South 24 Parganas, others deny displacement at all and claim that they have lived here all along. Jaya, by contrast, is from Uttar Pradesh (UP). Her husband used to be a truck driver, ferrying goods across the country, until he met with an accident and lost one leg. Since then, Jaya has been the sole earner for the family. In her own words: 'He had taken care of me all these years, now I would have to take care of him, right?' The other reason for the relatively more derelict shanty that Jaya and her husband inhabit is that the household is solely funded by Jaya's earnings. The migration from UP to Haldia is not something Jaya wants to discuss; however, sifting from other similar narratives, it seems likely that a belief in the 'growth' and industrialization of Haldia had encouraged many, including her and her husband, to make the move to settle here.

Women's migration, researchers hold, has to be read differently from male migration, even when the family migrates in its entirety (Pedraza 1991). Labour provides one of the bases on which this difference can be mapped. While men get involved in other forms of labour, especially in the case of rural to urban migration, women's labour continues in the form of cooking, cleaning, washing, and care work (Parrenas 2015). Whatever new identity as a participant in the labour force they might seek, be forced to adopt, or be denied, domestic labour continues. Along with this, as Ruchira Banerjee-Scrase and Kuntala Lahiri-Dutt (2012) note, migration to urban spaces usually means lack of access to the commons—forests and fields that were sources of fodder, fibre, and firewood in the villages (Banerjee-Scrase and Lahiri-Dutt 2012, 16–17). But in Haldia, as I outline below, the commons have been replaced by a set of common resources that are the result of the city's logistical structure.

Jaya's family survives on the making and selling of two kinds of *gul*, aside from any other possible sources of income that she omitted to mention. The word *gul* does not have an English synonym, indicating perhaps the indigenous and very poor contexts within which it circulates. Roughly translating into 'ball made of coal dust,' I am going to refer to it as 'coal pellets.' Used as fuel in small household clay ovens as well as in small eating establishments, tea shops, and even in the canteen of Haldia Government College (where I used to work), a giant sack of these pellets, amounting to 40 kilos, sells for INR 300. In the early and late afternoon, women can be seen making the pellets, placing them on plastic mats or torn tarpaulin sheets to dry during the day, gathering them up in the evening, and stacking them up for further drying the next day.

The manufacture and selling of coal pellets enables Jaya to identify herself as a wage earner, as one involved in financial transactions by virtue of her labour. But a closer look at the economy and ideology surrounding this alternative form of logistics would undermine her self-identification as an economic actor. The coal is gathered from two sources. Coking coal (used in steel production) and coal (used in power generation) are two main types of cargo handled at HDC. They are received at berths 4B and 8 which have an annual handling capacity of 2 million metric tonnes and 1.8 metric tonnes, respectively. In August 2015, the KoPT published a feasibility report for the construction of a new cargo-handling facility, a jetty at Outer Terminal II, specifically for coal. In the feasibility report, HDC is described as the preferred destination for importers of both coal and coking coal, and this cargo, once offloaded, is transferred onto rail wagons and trucks.

To the left of the entrance gate to the Dock Complex, between the wall that separates the port from uninhabited land and the rail line on which the coal-carrying wagons run, there is a drainage ditch that opens into a shallow, yet large, pond. Half of this pond is covered in water hyacinths, but the other half has been sectioned off using latticed bamboo partitions. The coal dust and sediment coming out through the drainage ditch gathers here; in monsoon, the sedimentation is higher. Women, like Jaya, get into the water to enhance the sedimentation process with their feet, gathering the sedimented coal dust with their hands and then carrying it home, leaving it to dry in mounds along the highway. Once the sediment dries a bit, the task of making coal pellets begins. The women make them into balls, dry them, and then pack them in sacks.

Coal is also sourced by women when the load-bearing wagons stop in the middle of fields, waiting for the green signal. Women climb onto the wagons and throw chunks of coal down to the ground, jumping back down as the train starts to move. Collecting the chunks, they go home to break them into coal dust and start the pellet preparation (Image 5.1).

Port logistics contributes to the making of another kind of coal pellets as well. When trucks carrying coal cargo stop at traffic signals on NH41, women from shanties run up to the trucks and start brushing the dust accumulated on the tires and various crevices of the vehicle into utensils and buckets. When the traffic is slow, the highway is swept by the same women, the dust gathered and sifted to isolate the coal particles. This also then produces coal pellets, which sell at the same price. The only difference between the first kind, gathered from sediments, and the second kind, gathered from sweeping, is that the second kind produces less heat (Image 5.2).

In third-world countries, the collection and preparation of fuel, alongside cooking, is almost exclusively a gendered task that is carried out by women of all ages. In rare instances, old men, who are no longer able to

Image 5.1 Woman throwing a chunk of coal down from a rail wagon that has stopped at a red signal, 2015. (Photograph by Samata Biswas)

Image 5.2 Left: Women scraping coal dust off coal-carrying trucks into cooking utensils as they wait at the level crossing, 2015. Right: Women carrying the gathered coal dust across NH41 in a bucket, 2015. (Photographs by Samata Biswas)

participate in remunerative labour, also contribute. It has also been noted that when fuel becomes more commercialized and technology driven, men's participation in this business increases (Dankelman and Davidson 1998).

Fuel is an everyday reality and requirement at Haldia. Apart from the Haldia Thermal Power Plant, which uses coal to generate electricity, other industries in the area also require large quantities of fossil fuel. In fact, coal forms not only the key component of Haldia's famous hinterland, consumption of coal by industries in Haldia itself merits, as mentioned above, three dedicated terminals at HDC. This coal needs to be loaded onto wagons and trucks, the trains and trucks have to be driven, and the coal has to be unloaded—all economically viable tasks that are undertaken by men. However, when used for household purposes, the procurement of coal becomes a woman's task. Then the coal's remunerative potential is removed and it becomes part of unpaid domestic labour.

In the case of these coal pellets, however, a certain very low form of income is generated by the women who collect the source of the fuel as a free by-product of the port's logistical operations. This process changes the ontology of the coal from one form of pollutant to another highly polluting fuel through the employment of sheer physical labour and then by its sale to food and tea stalls. Two crucial things needed to be noted in this case. Not only are the earnings from the selling of the coal pellets abysmally low, they are considered to be earnings merely because the labour involved in it is entirely women's labour, which is attached to other domestic work. In fact, the same fuel they sell at a price is also used by the women at home for cooking. But this labour can be done right at one's doorstep—in between cooking, cleaning, and care work. While women involved in this piecemeal industry continue to grow more abject in terms of time poverty (Warren 2003), their labour is necessary to feed the men who are value-generating parts of the existing logistical framework of Haldia, male labour that is increasingly undervalued, underpaid, and precarious.

Throughout the day, both beside NH41 and the Ring Road, one can spot women, children, and, at times, old men digging in the piles of mud used as landfill in the empty, low-lying pits alongside the roads. What they are looking for is iron scrap. One kilo of iron scrap brings in INR 400, but the amount of time it takes to generate one kilo is purely a matter of chance. It depends on the kind of clay or stones one dumper contains, and the scrap content within it (Image 5.3).

Scrap metal is one of the more important cargoes handled at HDC. In the financial year 2013–14, HDC handled 140,000 tonnes of scrap, and

Image 5.3 Women and children digging in landfill material for iron scrap, 2015. (Photograph by Samata Biswas)

in 2014–15, this went up to 156,000 tonnes (KoPT 2016). Once scrap is unloaded into a storage facility with the help of a Hydra, and the containers are emptied by tipping them over, men armed with shovels, hard hats, and gloves gather the scrap into mounds before loading them onto a truck or into another container. In contrast, the women and children who sort through the rubbish to find iron scrap, work with bare feet and hands. There is a logical extension to this scrap collection. Along both of the roads mentioned above, scrap-dealing shops have sprung up over the past few years. Scrap dealing, requiring greater investment of capital and yielding richer dividends, is a business engaged in solely by men, with women minding the shop only when the men are taking a nap. This scrap is then passed further down the chain, becoming part of big, male logistics.

HDC's logistical vision as well as the vision for Haldia as a smart city (an initiative of the Ministry of Urban Development of the Government of India, which seeks to improve overall infrastructure, provide sustainable real estate, and augment communications and market viability of selected

Indian cities) predicts a linear progress from agriculture to industry, from human labour to automation and machines. The interstices of dominant logistical frameworks that I have explored, however, point to a feminization of this 'other' logistics and to the presence of a labour force that does not enjoy basic rights and benefits. Keeping in place the key requirements of a logistical system: that is, being part of a supply chain network, and encompassing every step from the unloading of the cargo to its dispatching (as big logistical companies do), the women who pick up the excess, pollutants, and impurities of the cargo handled by HDC and its logistical counterparts—the railways, trucks, and warehouses—are neither bound by contract nor regularized, and they are ineligible for any kind of benefits. The intense labour involved in generating negligible value and the close association of this labour with domestic work foreclose the possibility of women emerging as viable economic actors in this alternative logistical network.

A desirable landfill material used in Haldia is the nonbiodegradable residue from the Kolaghat Thermal Power Plant 60 kilometres away. Haldia Energy Limited (a unit of Calcutta Electric Supply Corporation) also produces and sells landfill material and a truckload can be purchased for approximately INR 2000. It is then used like the clay in front of *Sonar Tari*—a proposed shopping mall in the city centre that has remained unfinished for ten years—to infill low-lying land next to the highway or the ring road and to fill the drainage ditches next to roads in order for various structures—from tailoring to butcher shops—to be built on them. Beside highways, where land stretches across the *nayanjuli* and agricultural fields, the filled-in space accommodates temporary shacks for vehicles or, more frequently, makeshift garages that repair large and small vehicles. Unlike Five Star Logistics, which has its own repair station, these repair shops are built on Municipality land with active help from the ruling party (whichever it might be). As they depend on the patronage of itinerant truck drivers for their subsistence, locations beside factory gates are the most favoured by these small businesses.

Conclusion

An act of othering, Simone de Beauvoir teaches us, is crucial to any self-definition, especially the self-definition of the hegemonic: 'Following Hegel, we find in consciousness itself a fundamental hostility towards every other consciousness; the subject can be posed only in being opposed – he sets himself up as the essential, as opposed to the other, the inessential, the object' (De Beauvoir 1989, xxiii). Big logistics, and the logistical vision that shapes and drives Haldia, thrives on a narrative of linear growth. The gaps in this narrative are filled by itinerant women workers like Jaya, who eke out their existence in a domain where male labour itself is increasingly more vulnerable, job contracts are short-lived, and the port, reported to be doomed to failure, continues with the business of logistical worlds.

While agricultural fields and fallow land may become entangled with other economies, transforming their value, but also transforming an already fragile ecosystem, relatively well-connected individual men may find a way out of poverty, and a life beyond agricultural or physical labour, even as their lives and employment remain increasingly entangled with that of the dock complex. At the same time, men's labour and the logistical side of the economy of the ports are subsidized by the labour of the women who cook and clean, maintain and are a key part of family units, and who find ways of boosting their income with their own resources and by dealing with the excess, the other, of logistics.

Acknowledgement I gratefully acknowledge the contributions of Suman Nath, Ananya Chatterjee, and Dipyaman Adhikary in carrying out the fieldwork. Uttam Chakraborty and Sk. Mujaffar were generous with their time, willingness to share experiences, information, and resources. Rupak Goswami helped immensely in the development of the argument. The participants in the Logistical Worlds project have listened patiently to the various versions of this paper and contributed to its present form.

Notes

1. Then called Calcutta, following the British name for West Bengal's capital, but in 2001, it was changed to the Bengali name, Kolkata. The Calcutta Port Trust (CPT) is now called the Kolkata Port Trust (KoPT).
2. ICARE runs Haldia Institute of Technology, Haldia Institute of Nursing Science, Dr B C Roy Memorial Hospital, Research and Development Centre, Vidyasagar Primary Teachers' Training Institute, Haldia School of Languages, Haldia Institute of Management, Haldia Institute of Health Science, Institute of Education, Haldia Law College, Global Institute of Science and Technology, Haldia Institute of Dental Science and Research, Haldia Institute of Maritime Studies and Research, and so on.

REFERENCES

Banerjee-Scrase, Ruchira, and Kuntala Lahiri-Dutt. 2012. *Rethinking Displacement: Asia Pacific Perspectives*. New York: Routledge.
Beauvoir, Simone de. 1989. *The Second Sex*. New York: Vintage Books.
Brandth, Berit. 1994. 'Changing Femininity: The Social Construction of Women Farmers in Norway.' *Sociology Review* 34, nos. 2–3: 127–49.
Brahmacari, Vankim. 1986. *Haldiar Itikatha*. Chaitanyapur: Midnapore Bharati Jana.
Dankelman, Irene and Joan Davidson. 1998. *Women and Environment in the Third World: Alliance for the Future*. London: Earthscan.
Deere, Carmen Diana. 2005. 'The Feminization of Agriculture? Economic Restructuring in Rural Latin America.' United Nations Research Institute for Social Development Occasional Paper. Geneva: UNRISD.
De Martino, Marcella, Luisa Erichiello, Alessandra Marasco, and Alfonso Morvillo. 2014. 'Logistics Innovation Networks for Ports' Sustainable Development: The Role of the Port Authority.' In *Sustainable Development in Shipping and Transport Logistics*, edited by Chin-Shan Lu et al., 435–44. Hong Kong: C. Y. Tung International Centre for Maritime Studies.
Economic and Political Weekly. 1976. 'Haldia: Still Born?' 11, no. 34: 1386–87.
Guha, Abhijit. 2008. 'Development Induced Displacement in West Bengal: Some Empirical Data and Policy Implications.' In *Status of Environment in West Bengal: A Citizen's Report*, edited by A. K. Ghosh. Kolkata: Society for Environment and Development.
Gupta, Jayanta, and Suman Mondall. 2016. 'Trinamool Not to Get Involved in Haldia Port's Affairs: Subhendu.' *The Times of India*. 15 February 15 2016. http://timesofindia.indiatimes.com/city/kolkata/Trinamool-not-to-get-involved-in-Haldia-ports-affairs-Subhendu/articleshowprint/50999726.cms
Grimshaw, Damian, and Jill Rubery. 2007. 'Undervaluing Women's Work.' Equal Opportunities Commission Working Papers Series. Manchester: Equal Opportunities Commission.
'Haldia City Census 2011 data.' *Census 2011*. http://www.census2011.co.in/census/city/252-haldia.html
Haldia Municipality. 2016. 'History of Haldia.' http://haldiamunicipality.org/new/public/about-history
HDA (Haldia Development Authority). 2016. 'Growth.' https://www.hda.gov.in/content/about-us&growth
Hill, Douglas, and Timothy J. Scrase. 2012. 'Neoliberal Development, Port Reform and Displacement: The Case of Kolkata and Haldia, West Bengal.' In *Rethinking Displacement: Asia Pacific Perspectives*, 31–52.

Kumar, Ujjwal. 2011. 'Study of Port Activities and Ship Scheduling Problem at Haldia Dock Complex.' MTech diss., Indian Institute of Technology, Kharagpur.
Kambouri, Nelli. 2014. 'Dockworker Masculinities.' In *Logistical Worlds: Infrastructure, Software, Labour*, edited by Brett Neilson and Ned Rossiter, 17–22.
Kolkata Port Trust. 2016. 'Traffic handled in terms of principal commodities at Kolkata port.' http://www.kolkataporttrust.gov.in/showfile.php?layout=2&lang=1&lid=1287
Nair, P. Thankappan. 1971. 'To Reharbour Calcutta.' *Economic and Political Weekly* 6, no. 27: 1327–28.
Neilson, Brett and Ned Rossiter. 2014. 'Logistical Worlds: Territorial Governance in Piraeus and the New Silk Road.' In *Logistical Worlds: Infrastructure, Software, Labour*, edited by Brett Neilson and Ned Rossiter, 4–10.
Parrenas, Rhacel. 2015. *Servants of Globalisation: Migration and Domestic Work*. Stanford: Stanford University Press.
Pedraza, Silvia. 1991. 'Women and Migration: The Social Consequences of Gender.' *Annual Review of Sociology* 17: 303–25.
Saha, Sambit, and Anshuman Phadikar. 2016. 'Anchor Role for Ripley in Haldia.' 30 March 2016. https://www.telegraphindia.com/1160216/jsp/business/story_69479.jsp
The Pioneer. 2014. 'CBI Raids Expose Haldia Port Scam, TMC Men in Dock.' 21 June 2014. http://www.dailypioneer.com/todays-newspaper/cbi-raids-expose-haldia-port-scam-tmc-men-in-dock.html
Warren, Tracey. 2003. 'Class and Gender Based Working Time? Time Poverty and the Division of Domestic Labour.' *Sociology* 37, no. 4: 733–52.

CHAPTER 6

Kolkata Port: Challenges of Geopolitics and Globalization

Subir Bhaumik

In his 1710 publication, *A New Account of the East Indies*, Alexander Hamilton wrote that, long before it became an 'imperial city,' Calcutta (now Kolkata) had 'docks for repairing and fitting ships' bottoms' (1995, 71). And Armenian traders possibly made use of these docks in their trade, with 'China to the East and Persia to the West,' before Job Charnock even set foot in the swampy villages of Kalikata, Sutanuti, and Gobindapur to lay the modern city's foundations. As Philip Woodruff remarks, it was from this port city that the British went on to create an empire 'at which they looked with incredulous elation, shot with sharp tinges of doubt, of a village grocer who had inherited a chain of department stores' (quoted in Moorhouse 1971, 37).

The Bhagirathi River, on which the Kolkata port developed, has been and remains a huge navigational challenge, but its location has never failed to entice the freebooter and the brave. A Dutch fleet of seven ships even negotiated the channel without pilots (as ships entering Kolkata rarely dare do) in 1759, two years after the Battle of Plassey, in a futile bid to

---

S. Bhaumik (✉)
Consulting Editor, Mizzima News, Myanmar and Former BBC Correspondent, Calcutta, India

deny the English the control of the Kolkata port. After the opening of the Suez Canal in 1869, the importance of the Kolkata port increased markedly—for both world trade and British interests.

The Muslim League stalwart and mastermind of the 1946 great Kolkata killings, H. S. Suhrawardy, unable to imagine an East Pakistan without Kolkata, pushed for its inclusion in the newly created country. India's current Prime Minister, Narendra Modi, has designated Kolkata as the 'starting point' of his 'Act East' policy for connecting India to South-East and East Asia. While Kolkata's port, a geomorphological nightmare, lies 232 kilometres inland from the sea, its multi-modal linkages to India's vast hinterland, as well as its eastern neighbourhood, have ensured its survival as an odd riverine port in an era of globalization that is now characterized by giant container-driven shipping.

The changing geopolitics of Asia, marked by China's sharp rise and India's emergence, has actually led to renewed interest in the Kolkata port system, amid ongoing questions about its future due to poor draught (water depth), ageing infrastructure, and the high cost of operations due to spiralling financial liabilities caused by its current large payroll, pension commitments, and the funds needed for regular dredging of the Hooghly channel. Nevertheless, the Kolkata Port Trust (KoPT), which handles the Kolkata and Haldia docks which make up the complex port system, is aware of its geopolitical and geo-economic importance, as this post on the KoPT website, by the former KoPT chairman R. P. S. Kahlon, shows:

> The process of churning a new and expansive trading hub, on a port-centric customer base with matching logistics and competitive facilities and tariff structure, while harnessing its riverine potentials, is one of the defining challenges the port faces in the years ahead. *The 'Look East Policy' of the country, the proposed Trans-Asian Railway corridor, opening of India-China road and proximity to Lhasa will all contribute to making Kolkata the hub port of the region.* (Kahlon 2016, my emphasis)

Indian policymakers may worry over possible Chinese maritime encirclement by a 'string of pearls' (China-constructed ports like Gwadar in Pakistan, Hambantota in Sri Lanka, Kyauk Pyu in Myanmar and, possibly, Sonadia in Bangladesh), but the KoPT chairman is indeed looking at a China-India road and reaching out to Tibet as possible options to augment the future business of the port. In fact, India's considerable diplomatic leverage in Nepal and Bhutan owes much to Kolkata port being the official port of the two countries. With all its economic muscle, China cannot provide a port in close proximity to Nepal and Bhutan, something that significantly adds to India's ability to influence politics in the two Himalayan nations.

The Chinese, for their part, have identified the Kunming-Kolkata (K2K) corridor, now popularly known as the proposed BCIM (Bangladesh-China-India-Myanmar) corridor, as one of six economic corridors to be developed under President Xi Jinping's Belt and Road Initiative. Chinese Vice-Premier Zhang Gaoli recently listed the six economic corridors as China-Mongolia-Russia, China-Central and Western Asia, China-Indo-China Peninsula, China-Pakistan, Bangladesh-China-India-Myanmar, and the New Eurasian Land Bridge which will be the focus of efforts related to Asia-Europe connectivity (Yang 2015). Billions of dollars in Chinese investment will flow into the corridors—which are designated to prop up trade and connectivity, infrastructure, and investment—and the Chinese-sponsored Asian Infrastructure Investment Bank will surely play a major role in their construction.

The BCIM corridor, which aims to connect Kolkata to Kunming in Yunnan province, is one of China's priority projects, although India seems to going slow on it. By getting India to declare Kunming and Kolkata as sister cities during a recent pairing exercise, China has made its priorities clear. Its limited coast in the East has compelled China to push for land-to-sea access in the Indian Ocean through both the Bay of Bengal and the Arabian Sea to avoid the Malacca Straits (which most Chinese security analysts see it as a 'chokepoint'). Besides Kyauk Pyu in Myanmar and possibly Sonadia in Bangladesh, Kolkata could serve as the third land-to-sea access in the Bay of Bengal for China but only if India agrees to operationalize the BCIM corridor.

If China's long-term goal is to use Kolkata as the third opening into the Bay of Bengal, this prospect has been somewhat mutely highlighted in an analysis:

> The Kunming-Kolkata corridor will not only rejuvenate the economies of south-west China, northern Myanmar, Bangladesh, east and northeast India, but will also connect to the proposed Amritsar-Kolkata growth corridor in India. Kolkata is thus the only Indian city that can connect a trans-regional corridor with an important Indian domestic growth corridor and the port system of the city will sustain a unique convergence of regionalization and globalisation. (Bhaumik 2014, 2)

No wonder, during the 2014 Kolkata-Kunming (K2K) Forum in Kunming, it was decided to 'develop several initiatives for greater connectivity between Kolkata and Kunming, with the BCIM economic corridor in mind' (K2K 2014, 2). But even if the K2K corridor does not materialize anytime soon, Kolkata will remain the 'starting point' of India's 'Look East' policy and its port system will be crucial for India's multi-modal connectivity to nations further east. With the coastal shipping agreement signed with Bangladesh during Modi's Dhaka visit in June 2015, India seems to have taken the first step in that direction. And the Kolkata port system—that will soon include a deep-sea component at Sagar Island and possibly another coastal port, besides Haldia, at Tajpur—will be crucial to connect India to its neighbours further east to South-East Asia and, if Delhi agrees, to China as well.

SAGAR: THE FUTURE OF KOLKATA PORT SYSTEM

Despite its obvious locational advantage, the Kolkata port has been hard put for survival. Draught in the Bhagirathi, the branch of the Ganges flowing into West Bengal, has come down over the years due to siltation. So much so that grants for dredging from the central government are seen as critical for the operability of the Kolkata port as the draught at Kolkata is never above 5 metres and, at Haldia, it is just above 6.5 metres. Kolkata and Haldia can accommodate Panamax ships (those that are capable of traversing the Panama Canal fully loaded), but the former KoPT chairman Bikram Sarkar and DHL Logistics Manager Deep Gupta both told me in personal interviews that even these ships have to reduce their load by half or more than half in other Indian ports or at the mouth of the river (from where they are transported by smaller vessels) to allow them to dock at Kolkata/Haldia port due to depth restrictions. As a result, Kolkata port's cargo volume dropped from an historic peak of 57.32 million tonnes in 2007–08 to 39.88 million tonnes in 2012–13. Through some local marketing moves and technological upgrades, the handling volume might pick up in future but only just a little. Atin K. Sen, former general secretary of the Asian Council of Logistics Services, indicates that large numbers of exporters and importers want to use Kolkata port for its splendid location, but they are discouraged by the rising costs that make its operations less competitive (Atin K. Sen, personal interview).

The huge silt volume brought down by the Ganges and deposited in its lower course impacts adversely on the Kolkata-Haldia port system. The development of the Haldia docks further down the river during the 1970s did provide some relief to Kolkata's British-era docks in the metropolis, but now the draught has begun to drop at Haldia as well, and this has necessitated the development of a true deep-sea port component to the Kolkata port system. Dredging has become key to the survival of the Kolkata-Haldia dock systems and the Indian government, which owns the port, has to provide between INR 3 billion a year for dredging the Bhagirathi to keep its channels just about navigable. The real beneficiary, needless to say, is the state-owned Dredging Corporation of India as the Kolkata port is its largest customer (Manoj 2013).

As the Indian government, headed by Prime Minister Narendra Modi, is considering phasing out most subsidies as a matter of policy, the dredging subsidy is likely to be slashed. According to Sanjoy Sen, former chairman of the Eastern India Shippers Association, however, the idea of using the funds saved by slashing the dredging subsidy to set up a new deep-sea port off Kolkata has been mooted by many since the late 1990s (Sanjoy Sen, personal interview).

In May 2013, the Indian government approved the setting up of a new deep-sea port at Sagar Island in the Bay of Bengal, near the confluence of the Bhagirathi and the sea. West Bengal's new Trinamul government lobbied hard with the central government on the issue, but, given the pace at which infrastructure projects are implemented in India, the port may take at least five to seven years to become operational.

At the same time and for similar reasons, in neighbouring Bangladesh the government is going ahead with the construction of a deep-sea port off Chittagong. Earlier, Sonadia was chosen as the location and China was supposed to fund its development, but now Matarbarhi has been chosen and Bangladesh is all set to construct it with assistance from Japan. Work was to begin in 2016 (bdnews24 2015), but the process of global tender for the port construction has been delayed, partly because of the uncertain security situation in Bangladesh after the terror strike in Dhaka in July 2016. Like Chittagong, Kolkata is a traditional port and experts say its handling volume cannot be improved merely by port modernization. Only a sprawling deep-sea component gifted with a substantial draught capable of handling large modern container-carrying vessels can reverse the situation. According to immediate past KoPT chairman, R. P. S. Kahlon, the successful integration of such a facility into the existing Kolkata-Haldia port system holds the key to the ultimate survival of India's only riverine port (R. P. S. Kahlon, personal interview).

The Indian government has already committed to invest the Indian rupee equivalent of USD 2 billion for the Sagar deep-sea project, and it has promised to pay back the West Bengal government which already invested INR 1700 million for the Sagar Island deep-sea port at Bhorsagar (*The Hindu* 2014). A study in 2014 by Assocham business chamber and the Institute of Chartered Accountants has indicated that a draught of at least 14 metres is needed for the Kolkata port system to receive big ships, without which its commercial viability is at stake (*Business Standard* 2014).

Neither the docks at Kolkata nor Haldia have the capacity for that kind of draught, even after extensive dredging. So just as Bangladesh is going for a deep-sea port at Matarbarhi to take the load off Chittagong, the only option for the Kolkata port system is to go for a deep-sea port, either at Sagar Island or another coastal port at Tajpur, that offers a draught of at least 12 metres.

After the 2016 West Bengal assembly elections, the Union government announced it will start work on the Sagar port by end of 2017. Nitin Gadkari, the Minister for Road Transport, Highways and Shipping in the central government, described Sagar as 'an addition to the 12 major ports of the country' (Gadkari 2016). According to the minister, the project will require an investment of the Indian rupee equivalent of USD 3 billion and that will involve a joint venture of the central government with the state government.

Now the central government has now made it clear it will only fund the Sagar port if the West Bengal government drops the idea of developing a port at Tajpur. The Bengal government, however, insists that it wants to develop the Tajpur port with private investments while the central government should finance the Sagar port (Saha and Sarkar 2016).

As well as the Sagar port project, Gadkari claims the Indian government is also constructing a rail-cum-road bridge, with the help of the National Highway Authority of India (NHAI), to facilitate inland transportation. He states that although Sagar will handle a significant portion of India's trade with Bangladesh after the two countries have signed a coastal shipping agreement, 95 per cent of India-Bangladesh's current trade, which is pegged at 6000 tonnes a year, takes place on land (Gadkari 2016).

> We have decided to promote waterways logistics. This will help in reducing the logistics cost (in India-Bangladesh trade) … The ministry of shipping and the ministry of external affairs are in talks with the Bangladesh government to have a port similar to Chabahar (in Iran). (Gadkari 2016)

India recently signed an agreement to develop Chabahar port for accessing Afghanistan and Central Asian countries and bypassing Pakistan. The Sagar-Paira port pairing in West Bengal and Bangladesh is seen as a similar geo-economic exercise by the Modi government to access the Indian northeast through Bangladesh by using waterways and providing a multi-modal option for making good the promise of the 'Look East' policy, which seeks to connect South-East Asia and India through the northeast. It also lends added importance to the Kolkata port system in which Sagar will constitute the deep-sea component.

There is a possibility that India Ports Global, a joint venture between Jawaharlal Nehru Port Trust and Kandla port, may participate in a bid to build the Paira port in Bangladesh, although it has also attracted China's interest. While Sagar and Paira will be paired as sister ports to promote India-Bangladesh coastal shipping more effectively, the Chinese see Paira as an ideal complement to their Sonadia deep-sea port in the Bay of Bengal and as a means by which China-Bangladesh trade can be boosted. China's Yunnan-Arakan oil-gas pipeline is complemented by a parallel highway that connects its largest province, Yunnan, to Myanmar's Kyauk Pyu port which was developed by the Chinese. According to Gadkari (2016), 'the shipping ministry is closely following the developments with regard to the Paira port so as to secure our national and strategic interests.'

The key to the whole exercise now is to strike a balance between the phasing in of cutbacks on the dredging subsidy for Kolkata-Haldia port and the contribution of sufficient funds to the timely implementation of the deep-sea project. But since the Modi government has also indicated a revamping of internal water transport in India, the dredging on the Ganges can never be totally cut down. Also if the cutback on the dredging subsidy is abrupt and immediate, it will impact on the existing business of the Kolkata Port Trust by causing a sharp drop in the handling of tonnage as bigger vessels give up on the port and instead turn to competitors like Paradip and Vizag on the east coast. The Indian government, therefore, may have to continue with much of the present level of dredging subsidy if it wants to keep the Kolkata port system alive and kicking until the Sagar deep-sea port is ready.

Nevertheless, Prime Minister Narendra Modi's government has a strong antisubsidy lobby led by Finance Minister Arun Jaitley. So while its move to set up the Sagar deep-sea port is a welcome development, and one that may save the Kolkata port system, what it does with the dredging subsidy is something to be closely watched. KoPT officials fear that once they lose out on handling volumes to competitors like Paradip and Vizag, it may not be easy to recover lost business, and then the Sagar deep-sea port is finally in operation.

As of now, the effort to revive Kolkata as a port system (rather than revive the riverside docks in Kolkata city) provides multiple options once the deep-sea component is ready. While the deep-sea port at Sagar will be able to handle large vessels and account for rising cargo volumes, Haldia will be able to compliment it by handling medium-sized vessels. Even the Kolkata docks can become a profit centre as a barge port for smaller specific cargo volumes and inland shipping (possibly extending to Bangladesh). The sprawling Garden Reach Shipbuilding Yard (GRSY) around the Kolkata port (whose orders are going up as India seeks to beef up its Navy and Coast Guard and also export ships) adds a profitable shipbuilding component due to increasing orders from the Indian Navy and Coast Guard. GRSY is also turning on a fairly aggressive campaign to export medium-sized patrol craft for Coast Guards in neighbouring countries like Bangladesh and Mauritius.

The KoPT also possesses swathes of what is turning out to be prime urban land. So far, some of it has been leased to companies like the filmmaking Shree Venkatesh Films, but the KoPT is drawing up a comprehensive privatization plan that will make the land available for real estate and some industrial renting. This is bound to boost KoPT coffers by adding to its kitty sizeable one-time funds and regular income from renting (*The Telegraph* 2015). A further source of income is the joint venture commercial partnerships KoPT is considering for riverside development to boost tourism.

But the deep-sea port at Sagar is also seen by some as only a medium-term solution. Logistics expert Atin K. Sen told me (in a personal interview) that a floating port at the Sandheads, further down from Sagar, could offer a lasting solution. Cargill on the west coast is a ready model for such an offshore port. It is based on an inventive port structure, designed to float offshore in deeper waters, that is capable of accommodating bigger ships. Cargill's massive platform debuted in India in 1998, and it features integrated cranes that unload more than 55,000 tonnes of dry bulk commodities while simultaneously loading other cargo for export. Floating five miles off the coast, goods are quickly unloaded from ships onto smaller barges and then ferried to shore for ground distribution (Cargill 2008). The Sandheads could be to the Bay of Bengal what Cargill is to the Arabian Sea. Its draught would be more than 50 metres at least.

KoPT reached its highest cargo handling volume in 2005–06 when it handled 41 million tonnes. That represented a rise of 35 per cent in two years, while all major ports grew by only 20 per cent in the same period. Only two major ports handled more cargo than Kolkata—the Kandla port at 41.5 million tonnes and the Vizag port at 47.7 million tonnes. In 2003–04, the Kolkata Port Trust clocked a growth of 15 per cent, compared to the 10 per cent growth recorded by all the major ports. Last year, container cargo grew by 19 per cent to about one million tonnes, behind only the Jawaharlal Nehru Port Trust and the Chennai Port Trust. Despite the spurt in cargo handling volumes, which was made possible by some management decisions like the abolition of the age-old tariff system and by cutting wharfage charges several times, the rise in handling volumes did not translate to greater profits.

In fact, between 2002–03 and 2004–05, there was a drop in operating profits by 16 per cent. Ever since then, KoPT's cargo handling volume has stagnated at between 35 and 42 million tonnes. In 2013–14, it handled 41.39 million tonnes against 39.28 million tonnes in 2012–13. The Haldia docks handled 28.51 million tonnes in 2013–14 against 11.84 million tonnes handled by Kolkata docks, clearly pointing to the limitations of the riverine port. Altogether, though, KoPT notched third place in handling container traffic. The number of vessels handled at Kolkata port during 2013–14 was the highest of all Indian major ports. During the year 2013–14, 3236 vessels called at KoPT, and it handled 17.1 per cent of the total number of vessels worked at Indian major ports in 2013–14.

For the Kolkata port management, tariff flexibility has not been the only strategy. It has tried to attract new types of cargo that would also give a boost to the economy of its hinterland. For instance, it found out that West Bengal had an annual surplus of 2.5 million tonnes of potatoes and that Singapore was ready to import more of this vegetable, which would otherwise either go waste or be sold in the domestic market at a throwaway price. While a few thousand container loads of potatoes had been exported, it was felt that such exports needed special containers and this increased the packaging costs. If only the port could reduce its cargo handling charges, Bengal's potato growers could reap huge export benefits. The Kolkata port authorities did not take long to reduce by half the wharfage charges for handling the potato containers. As a result, potato exports have become viable and the beneficiaries are the Kolkata port and of course the farm economy of Bengal (Bhattacharyya 2004).

Modernizing for Survival

The journey of the first major port of India—and its only riverine port—continues through the tortuous process of ebb and tide, expectancy and challenges, and is intricately woven with the varying draught/siltation challenges of the river Hooghly which cradles it. Nonetheless, the port continues to adapt and re-engineer itself in new locales and by diversifying its functions. The port of Kolkata has made investment decisions for both the Kolkata dock systems and Haldia dock systems. By keeping in mind the river morphology and traffic that can best be handled in these two locations, it is able to harness the potential of the two dock systems. A major challenge for a major riverine port like Kolkata has been its inadequate draught that prevents modern high-volume container ships from visiting the port. But being a riverine port and strategically connected to the National Waterway 1 and National Waterway 2, Kolkata port has huge potential in respect of the movement of cargo via the inland water transport (IWT) mode.

A well-diversified investment programme, aimed at improving infrastructure and augmenting capacity, is to be funded through internal resources and grant-in-aid by the government of India as part of its 11th/12th Five Year Plan. It encompasses construction of riverine terminals with improvement of back-end facilities as well as integrated development of infrastructure such as road/rail connectivity, improved storage yard logistics, induction of state-of-the-art equipment (such as stacker-cum-reclaimers), new Vessel Traffic Management (VTM) systems along with Automatic Identification Systems (AIS) with interfacing/integration, augmentation of IT infrastructure, and river regulatory measures for improvement of the draught at Hooghly estuary.

Apart from the above, various public-private partnership (PPP) and allied projects with an investment commitment of an India rupee equivalent of approximately USD 2 billion have been and are being taken up, the details of which are:

1. Major PPP projects included in 2012–13, spilling over to 2013–14
   - Development of berth facilities at Haldia Dock II (Salukkhali). There is a need for expansion of Haldia Dock Complex at an alternative location on the west bank of river Hooghly to cater to a higher volume of traffic, primarily coal, coking coal, and iron ore. Development of Haldia Dock II has been envisaged in the west

bank at Salukkhali/Rupnarayanchak, 15 kilometres northeast on the same bank from the existing HDC, where the land connectivity is conducive to cargo handling operations. Increased cargo throughput at the new site at a reduced overall logistic cost would be possible because of the availability of a relatively better draught, low turnaround time for the vessels by way of avoiding the existing lock system. Fresh Request for Quotation (RFQ) was issued in April 2012 Haldia Dock II (North) (cost INR 8 billion) and Haldia Dock II (South) (cost INR 8.8 billion), each having one mechanized and one multipurpose jetty, and with 23.4 tonnes—11.7 tonnes each—per annum capacity.

- Transloading facilities at Sandheads and its vicinity for midstream handling of dry bulk cargo

In order to overcome river draught problems, Kolkata port has already undertaken a string of initiatives aimed at setting up new port facilities at deep-draught locations. Priority has been assigned to the creation of transloading facilities within the limits of the port in an integrated manner with the construction of a riverine terminal (Outer Terminal 1) outside the lock gate at Haldia Dock. The transloading facility would avoid two port calls by mother vessels which are now visiting Haldia Dock with a much reduced load due to draught constraints.

- Public Private Partnership Appraisal Committee (PPPAC)

A meeting of this committee was held as early as July 2012. Earlier, activation of the project was subject to the outcome of the Special Leave Petitions (SLPs) filed by Kolkata Port Trust, Minister of State, and Government of West Bengal in the Honourable Supreme Court of India against the order passed by the Honourable Odisha High Court regarding the territorial jurisdiction of KoPT. During a recent hearing, the Honourable Supreme Court directed the parties concerned to have the matter amicably settled and action taken accordingly. Government approval has been obtained and as per observation of the Honourable Supreme Court, the locations for transloading operations have been identified with the neighbouring ports of Odisha. A draft notification prepared by the central Ministry of Shipping that declared the concerned locations to be within KoPT's limit is awaiting approval of the Government of Odisha.

2. PPP projects included in 2013–14 for placement of award
   • Diamond Harbour container terminal
      Development of a dedicated container terminal at Diamond Harbour, on the east bank of river Hooghly, was recommended by a high-powered committee set up by the Ministry. The project site, around 50 kilometres south of KoPT by road, is envisaged at an indicative cost of an Indian rupee equivalent of approximately USD 2.5 billion. The first phase of the project will comprise a contiguous quay length of 900 tonnes (a design capacity of 1.2 million TEUs). Projected container traffic is 1.2 million TEUs, that is, a 100 per cent rise in container handling is envisaged with the setting up of this container terminal—expected within four years from the date of award of contract. The first phase of the project will comprise a contiguous quay length of 900 tonnes including two jetties capable of handling container vessels of parcel load of 1200 TEUs.
3. Other major PPP projects to be taken up beyond 2013–14
   • Development of a deep-draught project for handling dry bulk cargo and containers at Sagar Island
      To establish port facilities at Sagar Island, including rail-road connectivity and construction of a rail-cum-road bridge over Muriganga, a feasibility study was entrusted to RITES Ltd who have since submitted the final report. In terms of decisions taken during a high-level meeting held recently at Kolkata, with a subsequent site visit, RITES made the necessary changes in the draft final report to project cost, implementation schedule, and economic analysis. Recently, a presentation on the proposed Sagar port was made before the Finance Minister of the Government of India by the Secretary of the Ministry of Shipping. The feasibility report submitted by RITES in December 2012, and approved in principle by the KoPT Board in January 2013, highlighted the need to set up a port facility at Sagar Island for handling 13.5 metres draughted vessels at a cost of INR 4.8 billion and allocation of another INR 3 billion for rail-road connectivity to enable the handling of 54 million tonnes of traffic in 2019–20. The project is currently awaiting cabinet clearance.
   • Development of Floating Storage Operations
      Owing to draught constraints, importers/exporters are unable to bring in fully laden vessels, and this leads to higher transporta-

tion costs. KoPT intends to address this by introducing Floating Storage Operations (FSO). Daughter vessels would transport liquid cargo between the oil terminals at Haldia and the floating storage. Tugs and other floating craft and equipment are required for undertaking Floating Storage Operations, and project parameters are to be soon finalized through initiation of feasibility study.

4. Schemes funded through grant-in-aid
   - River regulatory measures for improvement of channel draught
     River Regulatory Measures (RRMs) for improvement of the draught at Hooghly estuary, which comprise capital dredging and river training works, were earlier formulated by Hamburg University, and the scheme was approved by the Public Investment Board and subsequently by the Cabinet Committee on Economic Affairs (CCEA), at an estimated cost of INR 3 billion. The scheme could not be implemented as the bidders put in technical deviations in the form of unassured depth. On the backdrop of the earlier scheme, the entire project was reexamined by national and international experts and was finally revalidated by Water and Power Consultancy Services (WAPCOS), in association with Central Water and Power Research Station (CWPRS) and Lanka Hydraulic Institute, Sri Lanka. Due to a hike in the dredging cost and rapid siltation within the dredge-cut alignment, the entire project cost ballooned up to INR 1 billion. Since the unassured depth became the prime issue of contention and no one could assure a guaranteed result after implementation of the scheme, WAPCOS was requested to review the entire situation through an international expert. WAPCOS selected H. R. Wallingford, a renowned UK institute, for the said study. H. R. Wallingford, during a recent presentation to Minister of Shipping in March 2013, suggested KoPT not to implement the RRM in the present hydrodynamic condition of the river.

5. Projects aimed at opening the secondary channel and improving connectivity
   - Opening of Eden channel
     The existing system of pilotage through Eden channel was commenced on 1 December 2012. With the opening of the channel, ships enter Jellingham via Eden channel while bypassing the present Auckland channel. Steps are also being taken on the basis of recommendations of a high-powered committee to operationalize the channel throughout the year.

- Vessel Traffic Management System

    Kolkata Port Trust has installed a Vessel Traffic Management System (VTMS) with three radar stations at Frasergunj (S-band radar), Sagar (X-band radar), and Haldia (X-band radar) that are interfaced through a microwave link. The VTMS has worked satisfactorily since 1996 to provide effective navigational aid to vessels plying in the coverage area. To arrange the bare minimum necessary surveillance through instrumentation, installation of an Automatic Identification System (AIS) base station at Dadanpatra, as well as the introduction of a microwave link between Dadanpatra and Sagar, has facilitated the surveillance of vessels passing through Eden channel and its surroundings. A proposal for tender for design, development, installation, and commissioning of VTMS on a turnkey basis—at a cost of INR 330.5 million over a period of ten years, and inclusive of seven years of a central annual maintenance contract (CAMC)—has been taken up. It will involve replacing the entire system (Haldia, Sagar, and Frasergunj) with state-of-the-art technology, and all the four radar stations (Dadanpatra, Sagar, Frasergunj, and Haldia) will be equipped with 'X' band radar and dual transceivers through a microwave link.

Convinced that the Kolkata port system has a future in the emerging geo-economics of Asia, the West Bengal government has recently suggested another coastal port at Tajpur, not far from the fishing harbour of Shankarpur and the favoured tourist resort town of Digha in the state's East Midnapore district. The Bengal government claims that not only will the port at Tajpur be less costly than the one proposed at Sagar, it will also have greater draught which will allow large vessels with greater loads of cargo to dock.

While the draught at the deep-sea port at Sagar Island is expected to be 9 metres (as against Haldia's 6.5 metres), Tajpur could offer a draught of at least 12 metres. An estimate by the Bengal government indicates that the whole Tajpur construction project will cost no more INR 5–6 billion, while the one at Sagar will cost more than INR 12 billion. Sagar will also need much work on land reclamation and transport infrastructure to connect to the mainland, while those costs would be much less at Tajpur. The railways have estimated that at least INR 8 billion will be needed to connect the Sagar port to the existing railway system—and that includes the bridge over the Muriganga. West Bengal Chief Secretary Basudeb Banerji argues that while the biggest vessel that could dock at Sagar will not be heavier than 45,000 tonnes, the proposed port at Tajpur could accommodate ships with tonnage up to 60,000 tonnes (Basudeb Banerji, personal interview).

The Indian shipping ministry has said it will scrap the proposal for constructing the deep-sea port at Sagar Island if the Bengal government goes ahead with its plans to construct the port at Tajpur. Shipping ministry officials have told the media that in view of the current traffic and logistics trends, two ports so close to each other would not be viable in addition to the existing Kolkata-Haldia port complex (Saha and Sarkar 2016). In the meantime, India wants to approach the Japanese International Cooperation Agency (JICA) for a soft loan to develop the Sagar Island deep-sea port (or Tajpur).

REGION-MAKING: LOGISTICS OR GEOPOLITICS

Nations, either as part of regional groups or independently, seek alliances with like-minded countries to play a major role in the contemporary world. They create geopolitical regions that are often sustained by geo-economic drivers like trade and logistics. The Association for South-East Asian Nations (ASEAN) made sense in geo-economic and geopolitical terms when it was formed in the 1960s to bring together non-Communist nations of South-East Asia in a strong bloc that was capable, with US and Western backing, of staving off the challenge of communism. Its subsequent expansion to include the three communist regimes of Indo-China (Vietnam, Laos, and Cambodia) was not only made possible by geo-economic drivers like efforts to create a seamless production value chain but also by a felt need to neutralize China's rising economic power and influence. No wonder, Vietnam, once backed by China in its anti-colonial struggles against the French and later the US, is now strongly aligned to the US in its efforts to checkmate Chinese domination.

In South Asia, India failed to spearhead a comparable ASEAN-type grouping of South Asian nations, primarily due to its unrelenting rivalry with Pakistan, but it has now looked to its eastern frontier to create regional groupings as a means of forwarding its national interest. This includes the development of its underdeveloped east and northeast and the securing of its frontier zones, both militarily and economically. Much of India's trade has grown in recent years with South-East and East Asian countries. Since it is somewhat uncomfortable with the Bangladesh-China-India-Myanmar (BCIM) grouping because of China's presence and the dominant role it might play, India has joined two initiatives: the Bay of Bengal initiative BIMSTEC (which includes India, Myanmar, Thailand, Bangladesh, and Sri Lanka) and the new Bangladesh-Bhutan-India-Nepal (BBIN) initiative. The US and Japan remain strong allies, due to a mutual desire to contain China's rising influence and have shown ever-greater interest in strengthening relationships with both India and Bangladesh as well as the ASEAN nations.

Ports are emerging as important projects in the new chessboard of Asian geopolitics and geo-economics. To be economically viable, ports have to make geo-economic sense, but geo-economic viability in contemporary Asia is often driven by geopolitics. A huge Indian investment in Chabahar port in Iran appears justified because it would provide greater access to Central Asia and Afghanistan while allowing India to bypass Pakistan. Considering the area's huge energy resources, the money allocated by India to developing Chabahar appears well spent. If Pakistan had played ball, India would not need Chabahar. But since Pakistan is worried about India's rising influence in Afghanistan (and possibly Central Asia in the future)—and has shown a determination to oppose it, even to the extent of getting its intelligence agencies to use Taliban insurgents to attack Indian diplomatic establishments—India has to look to Iran and the port at Chabahar to achieve a crucial geopolitical and geo-economic objective. Chabahar is seen by India to have the potential to offset the China-funded port of Gwadar in Pakistan's Balochistan province, which is not far from Chabahar. That Gwadar lies at the coastal end of the proposed China-Pakistan Economic Corridor (CPEC) may be generating an Indian-US congruity of geopolitical and geostrategic interests.

In the east, India is now keen to further its bilateral relations with Bangladesh, not least to make up for its failure to normalize relations with Pakistan, despite some initial efforts by the Modi administration. Bangladesh is seen as crucial for Indian efforts to connect its mainland to the country's northeast, since the born-in-1971 nation sits between the two. In fact, developing close relations with Bangladesh is seen as crucial to furthering India's 'Look East' or 'Act East' foreign policy (Bhaumik 2014).

Integration of the transport infrastructure of India and Bangladesh to bring down logistical costs and facilitate a more seamless movement of goods and people seems to explain the Indian (and West Bengal) push for a deep-sea port at Sagar and/or a new coastal port at Tajpur, as well as the Bangladesh drive for deep-sea ports at Sonadia and Matarbarhi. But behind this lies the all-pervading Indian desire to connect closely to its own northeast through Bangladesh. A deep-sea port in Bangladesh's east that could reduce the load on the pre-British port of Chittagong is seen as crucial for India to access its northeast and connect further to South-East Asia. But if the deep-sea port is built by Japan, and not China, it suits not only Indian interests but also those of the US and Japan who want to contain China's rise and deny Beijing as many land-to-sea access points into the Indian Ocean as possible.

China seeks such access to avoid the Malacca 'chokepoint.' According to Chen Wenling, Chief Economist at the China Center for International Economic Exchanges, India's discreet but sustained lobbying in Bangladesh is responsible for the shelving of the Sonadia deep-sea port project (to have been constructed by China), and its replacement by the Japanese-funded Matarbhari deep-sea port project may not be unfounded (Chen Wenling, personal interview). India's interest in developing the Paira port in Bangladesh, and pairing it with a new port like Sagar/Tajpur in West Bengal, is seen as crucial to the success of the coastal shipping agreement between India and Bangladesh. This agreement aims to bypass logistical hubs like Colombo and Singapore in order to transport goods in large ships directly between Indian and Bangladesh ports. Such a plan makes logistical sense because it would bring down the cost of transporting goods between India and Bangladesh, not only because the bulk of mutual exports and imports would now be sent by sea instead of land transport through checkpoints like Benapole-Petrapole but also because such goods would no longer be carried to Singapore or Colombo as part of a great bulk and then shipped back. It would also cut down on transportation time and thus sharply increase bilateral India-Bangladesh trade which would strengthen bilateral relations in both economic and political terms.

A Bangladesh firmly integrated with India is seen as the first big step in Delhi's efforts to break the jinx in the East and develop regional groupings that offset the failure of similar efforts in the West while also containing or at least curbing Chinese influence. Much as Indian interest in Bangladesh developing a deep-sea port with Japanese rather than Chinese help is seen as crucial for its access to its own northeast and then beyond to South-East Asia, Indian investment in Sittwe port in Myanmar's Rakhine province as part of the Kaladan multi-modal transport project aims at both developing alternative access to its own northeast and connecting to South-East Asian nations through Myanmar.

The nearest Indian port to Sittwe is the Kolkata-Haldia complex. So it is not unexpected that India would be considering the future expansion of the Kolkata-Haldia port complex—either by furthering down the coast or by developing a deep-sea component to link up to ports in Bangladesh and Myanmar. Such an arrangement would further India's trade and commerce with two neighbours through whose territory it has to access its own northeast as well as the 'tiger economies' of South-East Asia and, if necessary, China.

Revitalization of the Kolkata port complex (including Haldia and future coastal/deep-sea port components) holds the key to Indian efforts at 'region-making' in eastern South Asia, which is itself a prelude for linking up to South-East Asia. Such efforts are seen by the US as bearing the potential for creating a strong bloc of rimland Asian nations capable of containing China's rise, which was undoubtedly the key objective of former US President Barack Obama's 'Asia rebalancing' strategy. Strengthening economic relations with friendly Asian nations, who are close allies of the US, was one of the declared objectives of Obama's 'Asia rebalancing' (Hall 2016).

Despite these developments, China sees ports like Kyauk Pyu in Myanmar or the proposed deep-sea port at Sonadia as capable of generating crucial land-to-sea access that holds the key to cutting down its dependence on the Malacca 'chokepoint' as well as generating potential huge savings in the logistical costs of imports and exports. Such access is also seen as crucial to China's current geo-economic and geopolitical vision that is expressed through the six proposed economic corridors as part of the Belt and Road Initiative. That vision seeks to create China's 'fraternal regions' through the expansion of key logistical infrastructure which will be made possible by generous funding from the China-sponsored Asian Infrastructure Investment Bank (AIIB). If India decides to play ball and goes ahead with the BCIM Economic Corridor, China sees the possibility of a third outlet to the Bay of Bengal-Indian Ocean region through Kolkata, as well as the proposed Sonadia deep-sea port in Bangladesh and the Kyauk Pyu port in Myanmar, which is also the starting point for an oil-gas pipeline to Yunnan. This may be one of the reasons why, as long as India remains engaged with it, the Chinese are taking the K2K Forum so seriously.

As geopolitical considerations influence Asia's emerging geo-economics and India joins China in the big league of nation-states, Kolkata and its port will grow in importance yet again. Not only does it serve a huge hinterland like the landlocked north-eastern states of India, the Himalayan countries of Nepal and Bhutan, and the Bihar-Eastern Uttar Pradesh region, it also sits at the centre of two proposed growth corridors—one national (Amritsar-Kolkata) and the other transnational (K2K) which could connect to eastern India and the neighbouring countries of Bangladesh, Myanmar, and South-West China. The Modi government has identified Kolkata as the 'starting point' of its 'Act East' thrust. But for the port system to take up these challenges, it has to survive. Modernization and expansion (with a deep-sea component and finally an offshore component) are seen to hold the key to the survival of the country's only riverine port.

REFERENCES

bdnews24. 2015. 'Matarbarhi Deep Sea Port Work to Start "Next Year".' 24 June 2015. http://bdnews24.com/bangladesh/2015/06/24/matarbarhi-deep-sea-port-work-to-start-next-year

Bhaumik, Subir. 2014. *Look East through Northeast: Challenges and Prospects for India.* Observer Research Foundation Occasional Paper 51. New Delhi.

Bhattacharyya, A. K. 2004. 'Why Calcutta port has a future.' *Business Standard*, 1 September 2004. http://www.rediff.com/money/2004/sep/01guest.htm

Business Standard. 2014. 'Kolkata Port Depth Needs to be Increased: Study.' 29 December 2014. http://www.business-standard.com/article/news-ians/kolkata-port-depth-needs-to-be-increased-study-114122900552_1.html

Cargill. 2008. 'The World's First Offshore Port.' Accessed 5 April 2017. http://150.cargill.com/150/en/ROZY-PORT-FLOATING-BARGE.jsp

Gadkari, Nitin. 2016. 'Coastal States Video Press Conference.' Press Information Bureau, Government of India video, 1:30:38. 3 June 2016. http://pib.nic.in/newsite/webcastplay.aspx?relid=0

Hall, Craig. 2016. Speech by US Consul-General in Calcutta at a book release function on 21 July 2016.

Hamilton, Alexander. 1995. *New Account of the East Indies.* Delhi and Chennai: Asian Educational Services.

Kolkata Kunming Forum (K2K). 2014. *Resolutions.* Yunnan Development Research Council (Kunming) and Centre for Studies in International Relations (Kolkata). 24 November 2014.

Kahlon, R. P. S. 2016. 'Message from the Chairman.' *Kolkata Port Trust.* http://www.kolkataporttrust.gov.in/index1.php?lang=1&level=0&lid=115&linkid=18

Manoj, P. 2013. 'All Above Board: Kolkata Port's Survival Blues.' *LiveMint*, May 17. http://www.livemint.com/Opinion/x9VN5Jp1xj1FnOaGXCbK6M/All-above-board-Kolkata-ports-survival-blues.html

Moorhouse, Geoffrey. 1971. *Calcutta: The City Revealed*. London: Weidenfeld and Nicholson.
Saha, Sambit and Pranesh Sarkar. 2016. 'Viability Cloud on Sagar Port Project.' *The Telegraph*, Kolkata edition, 5 August 2016. https://www.telegraphindia.com/1160805/jsp/bengal/story_100716.jsp
The Hindu. 2014. 'Gadkari Assures Support for Sea Port at Sagar Islands,' 24 December 2014. http://www.thehindu.com/news/cities/kolkata/gadkari-assures-support-for-sea-port-at-sagar-islands/article6721551.ece
The Telegraph. 2015. 'CPT Sniffs Big Bucks in Land Lease.' Kolkata Edition, 16 October 2015. http://www.telegraphindia.com/1151016/jsp/business/story_48267.jsp#.ViCpLuyqqko
Yang Ziman. 2015. 'Six Economic Corridors to Better Connect Asia and Europe.' *China Daily*, May 29. http://www.chinadaily.com.cn/business/2015-05/29/content_20858327.htm

PART II

Logistics of Asia-Led Globalization

CHAPTER 7

The Importance of Being Siliguri: Border Effect and the 'Untimely' City in North Bengal

Atig Ghosh

In 1951, Siliguri was an urban area with a population of just over six thousand in North Bengal, India (Ghosh et al. 1995). Within 65 years, however, the quaint half-town has been transformed into an enormous metropolis-in-the-making, and the pace of this transformation has accelerated in recent decades. Siliguri has become an important logistical crossroads, where the movements of people, arms, and other goods intersect. At once an entry point for contraband coming into India from China, and a hub in the hinterland of Kolkata port, the city's rapid expansion registers wider changes in politics and economy, and it is effectively redrawing the boundaries between traditionally recognized Asian sub-regions such as South Asia, Northeast India, South-East Asia, and Western China.

This chapter focuses on some key aspects or themes of this transformation—within the context of neoliberal accumulation from the 1990s—including defence, communication, migration, business, trafficking, and the tea plantation economy. These themes interlock and, further, the study

A. Ghosh (✉)
Department of History, Visva-Bharati (A Central University), Santiniketan, India
Calcutta Research Group, Kolkata, West Bengal, India

© The Author(s) 2018
B. Neilson et al. (eds.), *Logistical Asia*,
https://doi.org/10.1007/978-981-10-8333-4_7

does not approach Siliguri in isolation, as a monolithic self-completing urban phenomenon, or at a remove from the regional forces of economy and polity in which it is inevitably embedded and implicated and which it relentlessly shapes.

DEFENCE

Let me begin with an abbreviated history of Siliguri. The opening of the Darjeeling Himalayan Railway (DHR) in the 1880s imparted some importance to this burgeoning township as the place where 'the Corleones' of Kolkata culture—the Dasses, the Boses, and the Tagores—would break their journey to change trains for the hills. The tea trade that the DHR helped promote led to the expansion of the land and labour market in Siliguri and the establishment of Marwari *kothis* (settlements) in the area extended the informal capital and credit market. However, what transformed the scene radically was the partition of British India in 1947.

The formation of East Bengal, which was also known as East Pakistan and eventually renamed Bangladesh in 1971, created a geographical barrier in the north-eastern part of India. The narrow Siliguri Corridor—commonly known as the 'Chicken's Neck,' which at one point is less than 14 miles (22.53 kilometres) wide—remained as a national-territorial isthmus between the north-eastern part of India and the rest of the country. Siliguri thus found itself pitchforked into a position of immense geostrategic importance. Wedged between Bangladesh to the south and west and China to the north, Siliguri has no access to the sea closer than Kolkata, on the other side of the corridor, along the National Highways—NH31 and NH 34. Between Sikkim and Bhutan lies the Chumbi Valley, a dagger-shaped protrusion of Tibetan territory.

Consequently, there is massive military concentration in the area. The Siliguri Corridor is heavily patrolled by the Indian Army, the Assam Rifles, the Border Security Force (BSF), and the West Bengal Police. The North Bengal Frontier BSF is headquartered at Kadamtala, and one of the five frontier headquarters of the Sashastra Seema Bal (SSB) is located in Siliguri. These are two of the five Central Armed Police Forces of India, while the largest of the five, the Central Reserve Police Force (CRPF), also has considerable presence in the area. Two Air Force bases of the Eastern Air Command—the Hasimara Air Force Station (AFS) and the Bagdogra AFS—are located here which, for all practical purposes, are almost within Siliguri. The second largest military camp of Asia, Binnaguri, is also located not very far from the town. And if one were to assume that this massive concentration has allayed defence neurosis, one would not be more mistaken. On 19 November 2013, the army formally met with the West Bengal government to seek land to set up two military and air force stations in North Bengal 'to fortify the country's defence in the eastern sector' (*The Indian Express* 2013). The army officials, during the annual civil-military liaison meeting at Nabanna in Howrah, asked the state government for 750 acres of land at Dandim in Jalpaiguri for yet another Air Force station and 1000 acres in Kalimpong for a military station. Three months later, on 21 February 2014, the foundation stone for the Berhampore Military Station (BMS) was laid by President Pranab Mukherjee in the district of Murshidabad. At the ceremony, the then Army Chief General Bikram Singh informed the media that the BMS will be home to an Air Defence (AD) regiment where air defence missiles will be kept ready to protect the airspace over the Siliguri Corridor that connects the north-eastern states to the rest of the country (*Times of India* 2014).

Accounts of Siliguri often ignore this aspect of securitization, and, no doubt, the state probably wants us to ignore it. The more discussed aspects of Siliguri's history are the fairy-tale saga of its exponential growth and the rapid development of its local economy which, of course, cannot be understood fully without a reference to the presence of the Indian army, the BSF, the SSB, the CRPF, and the Assam Rifles around the city. Nevertheless, there is much more to the city's growth than this.

Pathways

Dubbed the gateway to Bhutan, Nepal, Bangladesh, and Northeast India, Siliguri stands in a unique geographical niche: Nepal lies in the west of the city 10 kilometres from Bagdogra, Bhutan to the northeast about 40 kilometres, and China to the north about 180 kilometres at Nathu La in Sikkim. To the south these days, Siliguri touches Bangladesh at Phulbari. Siliguri's strategic location makes it a base for essential supplies to the above regions and, as a result, the city has developed as a profitable centre for a variety of businesses including the so-called four Ts—tea, timber, tourism, and transport—which have long claimed to be its main businesses. While timber is languishing and tea has entered a complex career, however, a fifth T has ominously reared its head in the area: trafficking. Tea and trafficking will be discussed at some length later in the chapter. For now, I turn to transport.

As a gesture of international cooperation and friendship, the road network of Siliguri is being used by the governments of Nepal and Bangladesh to facilitate easy transportation of essential commodities, such as food grains. The Silk Route of India—that is, the trade route between India and Tibet (China)—is accessible through Nathu La and Jelep La only after crossing Siliguri (World Public Library 2016). And then, there are also the business routes to Bhutan across the Jaigaon-Phuentsholing border. Siliguri Junction, Siliguri Town, and New Jalpaiguri (NJP) are the three important railway stations within the Siliguri urban agglomerate. The Town station is the oldest. Started on 23 August 1880 by the colonial government, this station long served as the terminus for the trains coming from Kolkata and as the starting point of the internationally acclaimed DHR. The Siliguri Junction station was opened after independence in 1949 and used to be the point of departure of all trains to the northeastern states. To this station was added the NJP station in 1964. Located initially in Jalpaiguri district at a distance of 2.5 kilometres from the Town station, this was a wholly greenfield project. It is now the centre of rail communication for the entire region as well as Siliguri and, probably for this reason, has been incorporated recently into the Siliguri Municipal Corporation as one of its wards (17 of its 47 wards in the Jalpaiguri district). This is the largest railway station of the entire northeast and is, as we shall see, the nodal distribution centre for the trafficking flows from the region. The Bagdogra and Naxalbari stations may not be integrally a part of this network, but the former, which is located roughly 10 kilometres from the city, connects the airport there to Siliguri, while the latter, situated to the west of the Greater Siliguri City, is of immense strategic importance as it facilitates the people of Naxalbari and Panitanki to connect with other parts of the country and facilitates the people of Nepal from places like Kakarvitta, Dhulabari, and Bittamore across the Mechi River to utilize the railway station as the means of communication with the rest of India.

In recent times, the roads of Siliguri have assumed global logistical heft with the coming of the Asian Highways. The NH31 starts at Barhi in Jharkhand and passes through Siliguri to ultimately end at Guwahati, the capital of Assam. It has two tributaries so to speak: the NH31A stretches from Gangtok, the capital of Sikkim, to Sevoke on NH31, and the NH31C connects Galgalia in West Bengal to Bijni in Assam, skirting Siliguri at the Naxalbari-Bagdogra stretch. The importance of the NH31A can hardly be overstated as it is Sikkim's only substantial supply link from India. Cutting it off, as actions of the Gorkhaland separatist movement have often threatened in the Darjeeling hills, can cause severe distress to the northern state (*India Today* 2009; *Hindustan Times* 2011). The NH34 which meets the NH31 at Dalkhola in Uttar Dinajpur district begins at Dum Dum in Kolkata, thereby connecting Siliguri to the state's capital. Add to this the West Bengal State Highways 12 (Galgalia-Alipurduar) and 12A (Siliguri-Alipurduar), and we begin to develop a picture of the nodal location of Siliguri in a logistical road map that ties together Northeast India and the rest of India but also Sikkim, Nepal, Bhutan, and China.

The commencement of a major infrastructure project in the region, the Asian Highway (AH), to improve the highway systems in Asia, is a cooperative project between countries in Asia and Europe and the United Nations Economic and Social Commission for Asia and the Pacific (ESCAP). It is one of the three pillars of the Asian Land Transport Infrastructure Development (ALTID) project, endorsed by ESCAP at its 48th session in 1992, comprising the Asian Highway, the Trans-Asian Railway (TAR), and facilitation of land transport projects. Everywhere, in and around Siliguri, one notices the tracks that have been created for the multi-lane Asian Highways. Once completed, the AH2 will connect Panitanki near Siliguri on the India-Nepal border with Phulbari in the same district on the India-Bangladesh border, and the AH48 will connect Changrabandha on the India-Bangladesh border in Cooch Behar district with Jaigaon on the Jalpaiguri-Bhutan border (*The Hindu* 2014). These highways are to become, along with the network of rail and road already in operation, the lifelines of India's 'Act East' policy.

However, while we wait for these highway construction programmes to flesh out properly, which might put Siliguri in the shade, at least relatively, in terms of logistical importance, the pathways of capital seem to extend and ramify through the city's organizing heart to forge routes—probably not to the northeast—but beyond, to South-East Asian countries and China. Although it is too soon to feel excessively pessimistic about these developments, a caveat may not be out of place, that is, that they are likely to heed the tendency of neoliberal development to mobilize capital from point to point without any benefit for the lives of the populations it traverses en route. A literal, visual representation of this is a metropolitan flyover—say, the one in Siliguri—which leaps over the bazaar economy, urban squalor, and under-bridge homelessness of the city's most marginalized peoples in smug disarticulation with the economy of need. One wonders if 'Act East' will mean the same for Siliguri and the north-eastern states, where development will simply facilitate the movement of capital from the Indian mainland to South-East Asia, and leave the peoples of these regions in much the same historical predicament, poverty, as is suffered presently by those living in the under-bridge slums.

Migrant Business

As per the provisional reports of the 2011 census, the population of the city of Siliguri in 2011 was 513,264, including 263,702 males and 249,562 females. The Siliguri Metropolitan area (which includes Binnaguri, Chakiabhita, Dabgram, Kalkut, *and* Siliguri) has a population of 705,579, however, of which 362,523 are males and 343,056 are females. Hindus form the substantial majority of the population of Siliguri city at 91.98 per cent, followed by the Muslims at 5.37 per cent. According to the census, Siliguri is an urban agglomeration coming under the category of Class I UAs/Towns (Siliguri City Census 2011; Siliguri Urban Region 2011). In 1961, the population in Siliguri Municipal Area was 65,000, and it has increased by approximately 51 per cent every decade until it reached 227,000 in 1991 (Ghosh et al. 1995). If we take the statistics for Siliguri city alone, the population more than doubled once more from 1991 to 2011. Thus, while it is true that Siliguri has a long way to go before it can vie with the other urban behemoths of India, we have to take cognizance of this phenomenal growth rate. Very few cities compare to Siliguri in this respect and, moreover, calculations of the growth in number of 'residents' invokes an impression of sedentary urbanism that is probably unsustainable.

One commentator has gone so far as to call Siliguri a 'town in transit with the implication that it is the city that moves with its moving population and loses fast its potential of becoming anyone's home conventionally understood as the relatively stable abode where the family lives as like what Hegel calls "an individual"' (Das 2013). The transitional and transitory nature of this migrant city is not an effect merely of its fugacious daytime workforce. It is also an affect produced by the historical fact that waves of migrants have over the decades found a home in Siliguri, be they the Marwari settlers, old and new, the plantation workers of the nineteenth century, the partition refugees of the mid-twentieth century, or the recent flows of 'multi-collared' labour drawn by the lure of neoliberal lucre. 'According to a sample survey conducted in 1990,' writes Samir Kumar Das, 'amongst the immigrants, 60 per cent come from East Pakistan/ Bangladesh, while 17 per cent come from Bihar and 8 per cent happen to be Marwaris mainly controlling the wholesale trade. The remaining 15 per cent come from South Bengal or Assam' (Das 2013).

A comparative study of Baroda, Bhilwara, Sambalpur, and Siliguri—with reference to basic services for the urban poor—that was conducted in 1990 and published in 1995, provides figures for the migrants living in the slums of Siliguri (Ghosh et al. 1995, 210–215). According to the survey, 21.57 per cent of the total population of Siliguri resided in the slums of the city in 1991 (93). Today, going by the census of 2011, a population of 122,958 persons now resides in a total of 26,619 slums in Siliguri city. This is around 23.96 per cent of its total population (Siliguri City Census 2011). Given that the population has more than doubled in the last two decades, it can be claimed that the number of people living in slums has increased at an even faster rate and in even greater numbers. Faced with these statistics, works which *celebrate* the migrant nature of Siliguri (or any other city for that matter), the constant flux of its populace, the spatial tectonics of its geo-imagination, and the offering up of such unthinking descriptions of Siliguri as 'a cosmopolitan town in letter and spirit' (Chattopadhyay 1997, 48) can only elicit embarrassment.

At the other end of the migrant spectrum are the Marwaris, who control the wholesale trade in Siliguri. The Darjeeling district, in which Siliguri is mainly located today, can be said to have achieved its present shape and size relatively recently following the Treaty of Sinchula on 11 November 1865 between British India and the Kingdom of Bhutan (see Phuntsho 2013). It was, therefore, in the 1870s that the Marwaris first came to Siliguri (Saha 1993, A-345). From the late nineteenth century, the number of Marwaris in Siliguri continued to swell, though they preferred to live within the city and not settle beyond the town area, at least until the 1940s (Dash 1947, 71). In an undated interview conducted by Narayan Chandra Saha with Ram Kumar Agarwal, an old Marwari living in Siliguri, we have a guesstimate that about 30,000 Marwaris, of whom 5000 were Jains, lived in Siliguri at the start of the millennium (Saha 2003, 157).

Migration of the Marwaris to Siliguri did not follow a fixed pattern historically. In fact, it happened in four phases: pre-independence, post-independence, during trouble in Assam in the 1970s, and during the Bangladesh Liberation War in 1971. Even so, despite the relatively small size of the community, they have come to wield immense control over trade and commerce in Siliguri. For the longest period of time, they have been directly or indirectly connected to all kinds of enterprises—such as moneylending, *jotedari* (holding of *jotes*, which are a particular kind of land title), *aratdari* (big wholesale dealing), commission agencies, retail business, export and import business, ownership of hotels, restaurants, freight sheds and, lately, realty, tea plantations, and tea factories.

Given Siliguri's historical role as the coordinator of trans-border trade with Sikkim, Tibet, Bhutan, Nepal, and Bangladesh, as well as with Kolkata, it is not surprising that the nature of business in Siliguri is heavily biased in favour of wholesale and retail trade. It is basically a service town which, from the 1960s, has become an important wholesale trade centre. Much of this trade, as I have noted, is controlled by Marwaris, and strong kinship bonds among them make it practically impossible for others to freshly enter into the business. It is likely that these wholesalers possess a high amount of liquid cash, much of it derived as profit from their trade, which changes hands rapidly without getting anchored in investment to any great extent. The city also has a very high concentration of retail trade with an incredible number of shops in operation. There are three shops per 100 people, whereas in Delhi the number is 0.21. Many operate at a subsistence level and generate disguised unemployment in the informal service sector. The informal sector can expand to absorb unskilled labour—migrant labour from the slums—without any major capital expenditure. On the other hand, not uncommonly for India, at the owners' level, the retailer and the wholesaler are often one and the same person operating from the same shop front. Put another way, a bit of monetary inducement promptly turns the wholesaler into a retailer in Siliguri's bustling marketplaces.

Siliguri has experienced a very slow pace of industrialization. The number of industrial units fell from 174 to 162 between 1971 and 1985, though employment numbers remained stable at approximately 8300 workers (Census of India 1987). The combination of the trans-border trade system and lack of industry has made Siliguri 'a market town.' The Hong Kong market near Khudirampally is a chief hub for trading low-cost Chinese goods and imported goods; then there are the nearby Seth Srilal Market and the bazaars of Sevoke Road and Hill Cart Road. The Bagdogra airport market is one of the city's shopping hubs. Another important market is the Matigara *haat*, rumoured to be the oldest in the vicinity, where people from the hills and the plains come for shopping. In the past decade, the city has seen the establishment of a number of shopping malls and multiplexes. The organized retail sector has created jobs by employing unskilled informal labour, thereby extending the logic of disguised unemployment. Due to the city's cash-rich economic growth, many banks have also started to operate from Siliguri. As a prime service city, it has developed health and education infrastructure at an impressive scale, drawing people from neighbouring states and countries for treatment and education. This in turn has led to a boom in the hospitality industry. Siliguri has always had a greater share of hotels than similar towns because of the small traders who come and go incessantly. But, of late, the rate at which hotels in Siliguri have mushroomed along the axial roads is remarkable, belittled only perhaps by the number of shops and bars on these same roads. This tumultuous change, the flow of liquid cash, and the burgeoning culture of conspicuous consumerism have prompted the rapid growth of another 'service'—the flesh trade and the trafficking that grows out of it.

Trafficking

For most Siliguri residents, awareness of the trafficking racket in the city and its vicinity ranges from simple admission of its existence—without any deeper information or knowledge about it—to denial; and probably this is not unique, for most city dwellers, whether we speak of the Mumbaikar, Kolkatan, or Delhiwallah, conduct their lives without much knowledge about the seamier side of their urbanity. The lack of sensitization programmes alerting populations to the realities of trafficking may be blamed for such insouciance. Hence, questions about trafficking put to Siliguri residents are generally met with disconcerted mumbles.

The reality of trafficking is clearly a matter of anxiety for the police. The Siliguri Police Commissionerate (or Siliguri Metropolitan Police), established relatively recently in 2012 (Bhattacharya 2012), has a webpage dedicated to trafficking. Of the five commissionerates under West Bengal Police (the other four being Howrah, Bidhannagar, Barrackpore, and Asansol-Durgapur), only Siliguri has such a page. The way in which this page defines trafficking and categorizes it into four broad modes—humans, drugs, cattle, and ammunitions—shows both that Siliguri is under immense pressure to deal with trafficking and that it is a multi-modal problem. However, trafficking in persons (TiP) is the mode that is easier to grasp, and track, owing to the activism in the region related to this issue.

The Commissionerate regularly advertises its recent achievements in the 'Latest Update' section of its website. Of the self-advertised 'Good Work done in September, 2015,' there is one case that I present here to illustrate what the force considers a job well done and worthy of publicity. On 19 September 2015, Samandri Sahani of Matangini Hazra Colony lodged a complaint at the Siliguri Police Station (one of the six police stations under the commissionerate) claiming that her granddaughter, an 18-year-old named Aarati Kumar, had gone to a fair at about 8 pm but had not returned home. The 'complainant' had apparently conducted her own search to begin with and had approached the police only when she came to know that a man named Ganesh of the same locality had kidnapped her granddaughter and taken her to an 'unknown place.' The complainant further suspected, it was noted, that a man named Ranjit Paswan was also involved in this act, though this name was not mentioned in the police report. The police rightfully congratulated themselves on the fact that Aarati was rescued and the accused, Ganesh, was arrested the next day in Darbhanga district of Bihar (Siliguri Metropolitan Police 2015).

If you are thinking this could be a simple case of elopement and a spoilsport grandmother, you are terribly mistaken. It is surely a spoilt attempt at trafficking, though, not unlike an elopement, Aarati could have been an accomplice in her own trafficking; many acts of trafficking, in fact, begin as elopements at the source. The police records do not give us these details. However, Aarati's story follows a pattern that is now set in North Bengal. First, the teenage girl receives a missed call on her mobile phone. If she calls back, a male voice at the other end compliments the girl, say, on her voice or her wit. These 'phone relationships' then develop quickly with the stranger professing love for the girl and proposing to marry her immediately. This leads to that, and the impressionable teenager soon finds herself trafficked to strange cities and subjected to physical and sexual abuse.

Such TiP narratives suggest that Siliguri has become a centre for organizing flows of trafficked persons that originate in the city's vicinity (or beyond in the northeast) and end mostly in the metropolitan centres of India's north and west. This is substantially true. But the phenomenal growth of Siliguri has also meant that the city itself is often the destination for trafficked persons. In fact, over the last few years, many girls rescued from brothels in Delhi, Mumbai, and Siliguri have been from Assam (Rehman 2015).

Samir Kumar Das offers a more complex account of how trafficking fits into Siliguri's real-estate boom. He writes evocatively: 'The starred apartments of internationally mobile middle class and the new rich of Siliguri that remain vacant here reportedly serve as places of conduit where trafficked women – themselves in transit – are called to entertain their affluent customers in transit and money quickly changes hands' (Das 2013).

In this account, Siliguri is both a destination and a waiting-room, so to speak, of trafficked persons. However, since 2013, when Das wrote his piece, the situation has grown murkier with reports of what could be called 'reverse trafficking' becoming more prevalent.

In recent times, the richest recruiting ground for the traffickers in North Bengal has been the tea belt. The closed tea gardens probably provide the best opportunities for the traffickers, but gardens that are still operational are not exceptions. The heartbreaking fact is that in many cases prostitution or being trafficked is not the result of a person being duped into it; it is a conscious choice.

The closing of several tea estates in North Bengal, particularly those owned by the Duncan Group, has forced many young girls between 16 and 18 years of age into prostitution in order to feed their families. *The Times of India* (2016) reports a girl claiming: 'Some nights we even earn 500 rupees. The pimps take a portion of this but the rest is all ours. Who would have given us this money?' The statement makes any characterization of this predicament in clichéd terms of victimhood impossible and reveals the underbelly of the neoliberal development strategies affecting Siliguri. Another trafficked woman tells us: 'How long can we survive on government's mercy? We need many more things to survive than just rice. [The reference is to the state government's provision of rice at two rupees a kilo.] Now at least my family can have food every day; we have good clothes to wear and even have some spare money' (*Times of India* 2016). Debjit Datta, a prominent trade union leader, explains: 'The wages in the tea garden stand at about Rs. 122. This is also irregular in certain tea gardens. Don't you think that under such a situation women and girls would be more vulnerable to trafficking' (*Times of India* 2016)?

Datta is of course asking a rhetorical question. The suggestion is that if the wages and working conditions in the tea gardens of North Bengal do not improve, the flesh trade cannot be curbed nor trafficking ended. It is true that NGOs like Shakti Vahini are doing a lot to sensitize women in the tea gardens to the dangers of being trafficked as well as the strategies of traffickers (Shakti Vahini Press Release 2014). Success stories growing out of such sensitization programmes are not uncommon (Banerjee 2012). However, so long as life conditions force the 'victims' to collaborate, overtly or implicitly, in these processes, it is very difficult to put an end to them. The search for a solution, therefore, necessitates a consideration of the tea gardens of North Bengal, both those that have closed and those that remain operational.

Tea

A separate chapter would be required to do justice to the issue of tea in North Bengal. Even in the narrow context of Siliguri and its urban transformation, it is not enough to establish the link between the acquisition of tea garden land and the growth of prime realty around the city. There is more to the entangled story of the tea trade in Siliguri than a straightforward account of land grabbing and the realtor-politician nexus can deliver. However, I will start with this familiar theme and touch on some of the important issues towards the end. In 2005–06, 400 acres of land of the Chandmoni Tea Estate next to Siliguri were converted to real estate. Today the Uttorayon Phase I development, including the City Centre mall, stands in the place of these tea gardens. While a huge percentage of Siliguri's population languishes in slums, the municipal authorities are desperate to suppress this dystopia or at least overlay it with vision documents and dream projects for the developmental city. Hence, the official website declares: 'City Centre, Siliguri is a high point in the evolution of a growing metropolis that is increasingly influenced by four major drivers – quality housing, quality working, quality shopping and quality entertainment.' And why so? The mall operators answer: 'Siliguri has a vivid mix of local, transit and tourist population. City Centre will keep drawing visitors from both in and out of Siliguri – from adjacent foothills, the hills, the neighbouring areas, the borders and from across the City, providing a unique experience' (Siliguri City Centre 2016). In many ways, the developers have nailed it. They have got the character of Siliguri correct but not before cleansing it of the grime and sweat of poverty. The multicultural mélange on offer here is priced beyond the reach of even the urban lower middle class, let alone the tea garden workers who were evicted in the process of acquiring Chandmoni.

The brutal process of acquisition, under the slogan 'Siliguri is Changing,' has been written about in great detail. Samir Kumar Das (2013) gives us an effective and poignant account of the process. The General Manager of the Chandmoni Tea Estate presented the workers with a letter, giving them an ultimatum 'to vacate your present labour house, dismantle the same, vacate the premises and make it free from all encumbrances within October 31, 2005.' Incidentally, the letter is dated 5 January 2006, which makes it a *post facto* imposition of what Manas Dasgupta calls *forced voluntary retirement* on the tea workers; they had no possibility to negotiate their terms, let alone reject the proposal. The letter continues: 'You will be paid voluntary retirement compensation as per the said scheme with other terminal and statutory dues payable to you as per the terms and conditions of employment applicable to you' (Dasgupta, quoted in Das 2013, 83). Here we have, clearly, a case of primitive accumulation in the age of neoliberal development where tea labour is 'expropriated, evicted, pauperized, cut down from below the level of subsistence and thus pushed into hunger, penury and death through rampant use of violence and corruption' (Das 2013, 83). Dasgupta calls this 'Chandmoni capitalism,' but it is the common pattern of 'development' across North Bengal, the northeast, and, indeed, India at large (Dasgupta, 2006). Needless to say, the tea workers did not go gently into the dark night of casualization, deracination, and death; they fought a losing battle, and as early as 2002 two garden workers were killed by police gunfire. However, the continuing misery of workers was not always played out in mortal combat: many of them were ripped out of their social fabric and relocated to a distant tea garden in Subalbhita; many more were casualized and never rehabilitated, and many simply disappeared never to resurface.

This tale of the dispossession of tea gardens and their workers is a dismal narrative. What makes it even more unpalatable is that the conversion of Chandmoni into real estate was based on false claims. Dipankar Chatterjee, Managing Director of the Luxmi Tea Company, which bought the 'ailing' garden, makes this false claim: 'The Chandmoni Tea Estate was a loss-making company with huge liabilities. We could not turn the company into a profit-making one and have decided to convert 400 acres of tea garden into real estate' (*Business Standard* 2004). Chatterjee could be speaking for a whole range of tea garden owners who use the same pretext to sell gardens, convert them to other uses or, more commonly, keep the minimum wages of workers depressed. The first thing to note is that the tea industry is not ailing. If we look at the tea market in India, both supply and prices have increased in recent years. For instance, in 2012, 514.99 million kilograms of tea were produced and the average price in the market was INR 121.81; the following year production increased to 532.4 million kilograms, but instead of decreasing, the price increased to INR 128.46. Similarly, the auction price statistics of the Tea Board show that in 2006 and 2007, despite an increase in production, prices also increased.

There is nothing strange about these figures, because the demand for tea in India is far greater than its supply. According to the United Nations Food and Agriculture Organisation (UNFAO 2003), the rate of increase in domestic demand for tea was 3.7 per cent between 2003 and 2010, and to meet this demand 919.13 million kilograms of tea will have to be produced. Going by available statistics, India produced 966.40 million kilograms of tea in these years, of which it exported 222.2 million kilograms. Therefore, after export, only 744.2 million kilograms remained for domestic consumption. So there was a supply deficit of 175.1 million kilograms of tea. In this situation, it is unlikely that prices would fall and they did not. In fact, not only in the domestic market, but also in the international market the price of Indian tea has increased consistently. In 2007, tea sold for an average of 1.62 dollars per kilogram, whereas in 2010 the average export price of tea was 2.29 dollars per kilogram and in 2011 2.23 dollars per kilogram (International Tea Committee 2012). It also needs to be mentioned that the tea industry in India at present is not dependent on export and only a small percentage of the produce is exported every year. From 65 per cent of the total produce that used to be exported in the 1950s, it had climbed down to 19–20 per cent by 2005 (Karmakar and Banerjee 2005). This is because tea sellers have found the domestic market to be more attractively priced than the international market.

Clearly then, the planters' ploy of keeping wages depressed on account of an imminent fall in price of tea due to increasing supply and a collapse in export is disingenuous. In their report for 1888, the Indian Tea Association stressed the need to reduce expenditure 'by a more economic use of the labour available and by getting better work from the coolies' (Griffiths 1967, 123). Since tea was a labour-intensive industry, the colonial government clearly believed that production costs could be reduced by paying low wages to workers and also that a lesser number of workers could be squeezed to produce more. Sixty-five odd years of independence have not changed the planters' attitude much.

Today, in the sixteen big tea gardens owned by the largest corporate house in the Bengal tea sector, Duncan Industries Limited, permanent hunger stalks the lives of the 29,000 permanent workers and their dependents. These gardens have neither been abandoned nor declared closed by G. P. Goenka, the chairman of the Duncan Group, but left completely non-functioning, with the result that the workers cannot claim the INR 1500 per month government aid under the Financial Assistance to the Workers of Locked Out Industries scheme. Meanwhile, the powerful Goenka family that controls the Duncan Group has been siphoning off capital and considerable profit from tea to the company's sugar and fertilizer industries in Uttar Pradesh. They have done this by depriving tea garden workers of wages, bonuses, subsidized rations, fuel, and medical facilities; further, they have allegedly defalcated approximately INR 180 million from the workers' provident fund and gratuity dues. The same management has also cheated several frontline banks by borrowing huge amounts of money in the name of reviving 'ailing' gardens.

A similar situation obtains in other big gardens, which are lying abandoned or locked out by the speculator-owners (Mazumdar 2015). Ilina and Binayak Sen (2015), in a survey conducted at the Raipur Tea Estate near Jalpaiguri, found that of the 1272 workers surveyed, 539 (i.e., 42 per cent) had a body mass index (BMI) of less than 18.5. This is well above the 40 per cent critical value that is the baseline to categorize a community as famine-affected. While Mazumdar and the Sens report a permanent famine in the big tea gardens, both closed and operational, the planters have erected the bogeyman of ailing gardens to keep wages low and, when necessary, to sell them off. After such sales, new owners often step in with a bagful of promises, run the garden till they get a loan from the Tea Board or the banks to resuscitate the 'sick' garden, and then disappear with the money. It seems there is a booming market for *closed* tea gardens in North Bengal! The politician-realtor nexus that facilitates land grabbing in Siliguri's vicinity is but an extension of this process. In recent years, numerous tea garden workers have committed suicide or died of malnutrition and starvation. These deaths are organized institutional murders. Those who are selling themselves into prostitution and trafficking, too, are doing so because there is no discernible way to evade this structural violence and exploitation. While there are some positive stories about small tea growers who sell green tea to bought-leaf factories (Ghosal 2011), Siliguri remains the organizational centre of a plantation-based tea trade that perpetuates the exploitation, casualization, expropriation, and even death of tea workers (Bhattacharya 2015; Chaudhuri 2015). The city cannot escape the consequences of its involvement in this trade, for many casualized workers make a beeline for Siliguri to find jobs as day labourers, river-bed boulder-breakers, and construction workers, often in the very lands from which they or their brethren have been evicted.

Coda

In the above discussion, the idea of labour has been developed primarily through narrative examples. I close this chapter by offering a more theoretical explanation of my approach. Attentive readers will have grasped that I do not situate postcolonial labour *outside* processes of neoliberal development. Labouring subjectivities, in my view, are never encountered as *purely* resistant selves. At the same time, I do not want to frame labour as inescapably implicated in neoliberal developmentalism—as the co-constitutive double of logistical worlds, if you will. Labouring subjectivities are never *purely* docile bodies. I seek to understand how labour generates a dialectic of interiority and exteriority with respect to logistical worlds. This approach allows for a more historicized and dynamic understanding of labouring subjectivities in the context of Siliguri's metropolitan transformation.

Moving on, the political-economic *mélange* of people in flux—of wholesalers, retailers, traders, military and security personnel, tea planters and labourers, trafficked bodies and their consumers, gun-runners, political fugitives, asylum-seekers, railwaymen, construction workers, and stateless groups—cannot be understood outside the governing sign of neoliberalism. In fact, in 2002, India, Nepal, Bhutan, and Bangladesh considered a proposal to create a free trade zone in the Siliguri area which would allow the four countries to connect directly with each other without restrictions. Two observations are in order. First, it is not strange that an area which is a hub of heavy military securitization occasioned by neurotic obsession about territorial integrity is also a proposed hub of international free trade. This paradoxical coupling of military securitization and neoliberal circulation has been a marker of the northeast in general. Observing this paradox allows us to controversially state that Siliguri, and by extension North Bengal, is perhaps more integrally a part of India's northeast than it is a part of the rest of West Bengal. Second, the creation of a free zone for commodity trade is not awaiting an international diktat in the area. As has been suggested above, such a zone has already come into being informally. A walk through the bazaars of Siliguri testifies to this. The new shopping complexes have not even offered the meekest challenge to the booming business of the Seth Srilal Market, the Sevoke Road, and Hill Cart Road bazaars or the airport market at Bagdogra. These markets are prominent places to buy daily-use goods and are extremely popular among people from nearby areas as well as tourists from all over the world. They are awash with commodities imported from neighbouring countries, by means both licit and illicit.

Like so many other spaces in Siliguri, the bazaars dent the city's sanitized self-image as a hub of neoliberal development and reveal the seamier, probably dominant side of a border town in the throes of economic expansion. Siliguri is a node for smuggled goods and trafficked human beings. These are the logistical flows of crime and commerce that co-constitute Siliguri under the sign of neoliberal capital. They are flows that attract constant commentary in government reports and newspaper articles but are difficult to grasp and render in substantiated academic commentary. Yet, they impart an unmistakeable flavour to the sensorium of this metropolis-in-making. Historians have been arguing for the coevalness of colonial trade and administration with the bazaar economy in South Asia for some time now (see, for instance, Bayly 1983). Beyond the mere coexistence of two apparently incompatible modes of consumption, what seems to be at work is an active gerrymandering of collective and individual urban subjectivities. It is as if the rational actor-citizen has been entirely displaced by the migrant as the organizing human principle of political economy. Further, to borrow a term from cinema studies, the Siliguri cityscape presents itself as retro-futuristic, as depictions of the future produced in an earlier era. Here, the neoliberal sign appears coded in the migrant bazaar.

Through this compressed and somewhat schematic commentary on Siliguri's transformation, I have tried explore the possibility of certain conceptualizations. First, it seems that there is a need to rethink the geo-imagination of North Bengal at a time when its material and ideational coordinates are being expanded by statist defence neurosis as well as everyday practices of mobile peoples. Second, it is possible to understand this re-imagined North Bengal as an integral part of India's northeast, with its border economy and 'travelling actors.' Third, through entanglements of control, crime, communication, and capital, Siliguri shows us that the border economy does not remain confined to the border and borderlands but seeps into the so-called mainland to bring about powerful economic transformations. Going a step further, it may be said that the 'metro-polarities' of Siliguri present what can be called a 'futuristic archetype' of a border city. It is archetypical in the Jungian sense of being a mental image—a dream project—that is always already present in the collective unconscious and yet, insofar as it is a mental image, remains an abstraction only realizable at some indeterminate and permanently deferred point in the future.

References

Banerjee, Amitava. 'Student Busts Human Trafficking Racket in Darjeeling.' *Hindustan Times*, 22 March 2012. Accessed 20 March 2016. http://www.hindustantimes.com/kolkata/student-busts-human-trafficking-racket-in-darjeeling/story-2g27t89BlimLgOfbnLxzJO.html

Bayly, Christopher A. 1983. *Rulers, Townsmen and Bazaars: North Indian Society in the Age of British Expansion, 1770–1870*. Cambridge: Cambridge University Press.

Bhattacharya, Ananya. 2015. 'Black Tea: Shut Tea Estates and the Pied Pipers of North Bengal.' *DailyO*, 8 May 2015. Accessed 21 February 2016. http://www.dailyo.in/politics/black-tea-shut-estates-and-the-pied-pipers-of-north-bengal/story/1/3620.html

Bhattacharya, Pinak Priya. 2012. 'Mamata Banerjee Inaugurates Siliguri Police Hub.' *Times of India*, 6 August 2012. Accessed 1 March 2016. http://timesofindia.indiatimes.com/city/kolkata/Mamata-Banerjee-inaugurates-Siliguri-police-hub/articleshow/15370009.cms?referral=PM

Business Standard. 2004. 'Chandmoni Tea Uprooted for Siliguri's First Township.' 5 January 2004. Accessed 21 February 2016. http://www.business-standard.com/article/companies/chandmoni-tea-uprooted-for-siliguri-s-first-township-104010501098_1.html

Census of India. 1987. *District Census Handbook: Darjeeling*. Kolkata: Directorate of Census Operations, West Bengal.

Chattopadhyay, Sibaprasad. 1997. *Known Yet Unknown Darjeeling/Siliguri: Facts and Figures*. Siliguri: Kashi Nath Dey.

Chaudhuri, Mohuya. 2015. 'Tea Gardens in the East are Brewing Starvation, Malnutrition.' *The Wire*, 30 July 2015. Accessed 21 February 2016. http://thewire.in/2015/07/30/tea-gardens-in-the-east-are-brewing-starvation-malnutrition-7571/

Das, Samir Kumar. 2013. 'Homelessness at Home.' *Eastern Quarterly* 7, nos. 3–4.

Dasgupta, Manas 2006. *Uttarbange Chaa Shilpe Bartaman Shamashya* [The Present Problems in the Tea Industry of North Bengal]. Kolkata: Boiwallah.

Dash, Arthur Jules. 1947. *Bengal District Gazeteers: Darjeeling*. Alipore: Bengal Government Press.

Ghosal, Sutanuka. 2011. 'Bengal Entrepreneurs Look at Tea Production as a Source of Income.' *The Economic Times*, 27 December 2011. Accessed 21 February 2016. http://articles.economictimes.indiatimes.com/2011-12-27/news/30561684_1_tea-production-green-leaf-tea-growers

Ghosh, Archana, S. Sami Ahmad, and Shipra Maitra. 1995. *Basic Services for Urban Poor: A Study of Baroda, Bhilwara, Sambalpur and Siliguri*. Urban Studies Series No. 3. New Delhi: Institute of Social Sciences & Concept Publishing Company.

Griffiths, Percival. 1967. *The History of the Indian Tea Industry*. London: Weidenfeld & Nicolson.
The Hindu. 2014. 'Bengal Proposes Road Connecting State with Three Countries.' 4 June 2014. Accessed 20 February 2016. http://www.thehindu.com/news/cities/kolkata/bengal-proposes-road-connecting-state-with-three-countries/article6081600.ece
Hindustan Times. 2011. 'NH 31A Open for Sikkim Traffic: GJM.' 14 February 2011. Accessed 21 February 2016. http://www.hindustantimes.com/india/nh-31a-open-for-sikkim-traffic-gjm/story-EX4eiPX5CDZLyG855xTz3I.html
The Indian Express. 2013. 'The Army seeks Land to set up Two Stations in North Bengal.' 19 November 2013. Accessed 15 February 2016. http://archive.indianexpress.com/news/army-seeks-land-to-set-up-2-stations-in-north-bengal/1196625/
India Today. 2009. 'National Highway 31A Blocked by GJM, Sikkim Cut off.' 7 February 2009. Accessed 21 February 2016. http://indiatoday.intoday.in/story/National+Highway+31A+blocked+by+GJM,+Sikkim+cut+off/1/27551.html
International Tea Committee. 2012. *Annual Bulletin of Statistics*. London: International Tea Committee.
Karmakar, K. G. and G. D. Banerjee. 2005. *The Tea Industry in India: A Survey*. Mumbai: National Bank for Agriculture and Rural Development.
Mazumdar, Abhijeet. 2015. 'Hunger Valley.' *Newsmen*, 16 November 2015. Accessed 21 February 2016. http://www.newsmen.in/news-item/opinion-hunger-valley/
Phuntsho, Karma. 2013. *The History of Bhutan*. New Delhi: Random House India.
Rehman, Azera. 2015. 'Assam's Tea Gardens are Waging a War Against Girl Trafficking: It's both Inspiring and Heartbreaking.' *The Better India*, 29 August, 2015. http://www.thebetterindia.com/32118/assam-tea-gardens-girl-trafficking/. Accessed 20 March 2016.
Saha, Narayan Chandra. 2003. *The Marwari Community in Eastern India: A Historical Survey Focussing on North Bengal*. New Delhi: Decent Books.
Saha, Narayan Chandra. 1993. 'Darjeeling Tarai Anchaler Marwari Samaj: Ekti Samiksha [The Marwari Community in the Darjeeling-Terai Region: A Survey].' *Itihas Anusandhan* 8.
Sen, Ilina and Binayak Sen. 'Cup of Starvation.' *The Week*, 17 March, 2015. Accessed 21 February 2016. http://www.theweek.in/columns/Binayak-Sen/cup-of-starvation.html
Shakti Vahini Press Release. 'Child Trafficking – a Major Menace in West Bengal.' 26 August 2014. Accessed 20 March 2016. http://shaktivahini.org/press/child-trafficking-a-major-menace-in-west-bengal/

Siliguri City Census. 2011. *Census 2011.* Accessed 21 February 2016. http://www.census2011.co.in/census/city/192-siliguri.html
Siliguri City Centre. Accessed 21 February 2016. http://siliguri.citycentremalls.in/
Siliguri Metropolitan Police. 2015. Accessed 21 February 2016. 'Good Work Done.' https://www.siliguripolice.org/gwd
Siliguri Urban Region. 2011. *Census 2011.* Accessed 21 February 2016. http://www.census2011.co.in/census/metropolitan/186-siliguri.html
Times of India. 2014. 'Army starts Infrastructure Development in Bengal with Air Defence Station in Murshidabad.' 21 February 2014. Accessed 15 February 2016. http://timesofindia.indiatimes.com/city/kolkata/Army-starts-infrastructure-development-in-Bengal-with-air-defence-station-in-Murshidabad/articleshow/30771758.cms
Times of India. 2016. 'West Bengal Shut Tea Gardens Force Girls to Resort to Prostitution.' 5 February 2016. Accessed 20 March 2016. http://timesofindia.indiatimes.com/city/kolkata/In-West-Bengal-shut-tea-gardens-force-girls-to-resort-to-prostitution/articleshow/50865125.cms
United Nations Food and Agriculture Organisation (UNFAO). 2003. Medium-Term Prospects for Agricultural Commodities: Projections to the Year 2010. Rome: UNFAO.
World Public Library. 2016. 'Siliguri.' Accessed 20 February 2016. http://www.worldlibrary.org/articles/siliguri

CHAPTER 8

Piraeus Port as a Machinic Assemblage: Labour, Precarity, and Struggles

Pavlos Hatzopoulos and Nelli Kambouri

On 17 February 2016, COSCO Group Ltd was declared the preferred investor for the acquisition of 67 per cent of the shares of the Piraeus Port Authority (OLP) by the Hellenic Republic Asset Development Fund (HRADF). This announcement ended a prolonged period of uncertainty, following the formation of the newly elected coalition government of the left-wing party SYRIZA and the right-wing nationalist party ANEL in January 2015, which had initially declared that the ongoing 'privatization' of the Piraeus port and other critical infrastructures would be reverted. The privatization of Piraeus port was officially completed in August 2016. Through a concession agreement, the COSCO Group has been granted the management of all port services, including coastal and cruise shipping, the car terminal, and the remaining Pier I of the container terminal, which was still run by OLP. The concession will expire in 2052, prolonging COSCO's already privileged presence in Piraeus and making it practically a monopoly company in all services and facilities of the port.

P. Hatzopoulos
Open University of Cyprus, Athens, Greece

N. Kambouri (✉)
Panteion University, Athens, Greece

© The Author(s) 2018
B. Neilson et al. (eds.), *Logistical Asia*,
https://doi.org/10.1007/978-981-10-8333-4_8

This recent privatization of Piraeus port is in fact an extension of an earlier 2009 concession agreement signed between the Greek government and COSCO Pacific Ltd in 2009, which had radically transformed labour relations in the port. Our fieldwork research in the port of Piraeus started in 2013, at a period when the container port was divided into two parts: Pier I continued to be managed by the existing port authority (OLP), while Pier II and, from 2013, the newly constructed Pier III were operated by Piraeus Container Terminal SA (PCT), a COSCO subsidiary company operating under Greek law. Following the concession, there emerged in the port new machines, softwares, infrastructure, entrepreneurial cultures, and languages—some in material form and others in the form of anticipations, projections, and desires. At the time, we could only sketch what was emerging next to the existing OLP regime by juxtaposing conflicting statements announced by Greek government officials, COSCO executives, Chinese diplomats, Greek labour unionists, technicians, and inhabitants of the surrounding area. The amalgam of these enunciations, however, already sketched a complex assemblage of machines and humans that produced the port anew.

The prevailing neoliberal enunciations of the COSCO concession projected the futuristic scenarios of economic efficiency, the utilization of new technologies and highly trained personnel, and the undertaking of research and innovation on new 'infrastructures and superstructures' that would transform Piraeus into the 'largest port in Eastern Mediterranean' (PCT 2016). As Greece was entering a period of deep recession and austerity, government officials celebrated these progressive visions as the largest 'investment' on the Greek territory, emphasizing the prospects of further Chinese investments that would bring growth and jobs to the stagnant economy. According to the 'Chinification' thesis, however, the concession was portrayed as a step backwards in time—imposing primitive, oriental, and autocratic labour practices in the port—that was often compared to work in the Roman galleons or indentured labour in medieval Europe. Instead of growth and technological efficiency, critics argued that the COSCO concession was a form of 'Chinese colonialism' that would disperse the backwardness of its precarious labour regime outside the port, into the whole of the city of Piraeus and, eventually, throughout the entire Greek territory.

In this chapter, we argue that, after the COSCO concession, Piraeus has indeed been colonized, but not—or at least not only—by the Chinese, for as Donna Haraway has argued, '[m]odern production seems like a dream of cyborg colonization work, a dream that makes the nightmare of Taylorism seem idyllic' (1991, 292).

We view the port as a machinic assemblage that creates new possibilities and desires that extend beyond its spatial confines (Deleuze and Guattari 1999). In this, we draw on Deleuze and Guattari to argue that what was born in Piraeus after the concession was not only a new form of precarious 'alienated mechanized labour' but also a new set of social relations of 'men and women' becoming 'part of the machine not only in their work but even more so in their adjacent activities, in their leisure, in their loves' (1986, 81). Moreover, we use Haraway's 'blasphemous critique' of the prevailing male-dominated Western political tradition of thought to understand labour in Piraeus in terms of the borderline figure of the cyborg: 'a cybernetic organism, a hybrid of machine and organism, a creature of social reality as well as a creature of fiction' (1991, 291).

From this perspective, we argue that, in the container terminals of Piraeus, labour can no longer be confined to the representations of workers' bodies performing repetitive tasks in order to operationalize machines. This labour is enabled, it is argued, through the everyday operations of cybernetic organisms whose lives are ordered according to the objectives of the maximization of efficiency and the minimization of idleness. The COSCO concession, as Brett Neilson suggests, can be approached both in terms of the political technologies involved in the establishment of complex layers of territorial governance of the port and, concomitantly, as a site of logistical operations that utilize software systems to design and maximize efficiencies. In this respect, the intersections between these political and logistical technologies are critical for the analysis of the transformation of 'labour processes and relations' (Neilson 2018), which we will try to address in this chapter.

Our analysis of the transformation of labour processes and relations in the port of Piraeus is disjointed from the growing discussion of the internationalization of Chinese firms and its impact on domestic labour relations in Europe (see, for example, Smith and Zheng 2016; Burgoon and Raess 2014; Meunier 2012). On the one hand, we are very supportive of the voices that are critical of the association of Chinese investments with the imposition of a distinct, Chinese labour model that undermines existing European labour laws and norms. It is also along these lines, however, that we question simplistic readings of the globalization of Chinese expansion by highlighting the lack of a unified, integrated model of work organization and employment relations amongst Chinese firms (Zhu et al.

2012; Child 2009) and the lack of an integrated European space and European labour market given the segmentation of the so-called European labour regime and the current transformations of labour relations in Europe, especially in relation to processes of precarization. Additionally, there are some analyses that raise very important questions regarding the possible coevolution of labour relations, in China and overseas, and especially concerning the potentialities of connecting emerging labour struggles in China with labour struggles around the operations of Chinese firms overseas (Smith and Zheng 2016).

Our research, however, was primarily concerned with an investigation of how the COSCO concession of the Piraeus port has enabled the production of new labour subjectivities and is currently shaping processes of subjectivation and desubjectivation. Our focus on these processes brings to the forefront the conditioning of Chinese-led globalization by national and local labour regimes.

The first section of this chapter is an attempt to analyze the COSCO concession as a set of often conflicting processes of reterritorialization and deterritorialization that produce new labour subjectivities in the Piraeus port. The second part of the chapter relates to the analysis of the regimes of control of production and labour in the Piraeus container terminals. The final section analyzes the notion of idleness and how it becomes central to the development of processes of optimization of production.

PIRAEUS AS 'A TERRITORY WITHIN A TERRITORY'

On one of our first visits to the port, we were invited to join a guided tour of the pier that was still operated by the Piraeus Port Authority (OLP). The tour was organized by the OLP labour unions, and our guide was a highly educated left-wing female activist and trade unionist who was working in the cruise section that remains under the OLP administration. During the tour we were shown an invisible dividing line—a time-space border—that separated two distinct labour regimes. Although there was no physical barrier, one could not cross this line as entrance into the piers that were run by COSCO was strictly forbidden, especially for researchers. Similarly, a COSCO inspector searching for a lost container in Pier I couldn't just cross the invisible line either: they had to exit Pier II in order to reenter Pier I through the OLP security checks.

Our labour union guide pointed out repeatedly that OLP was constrained within a very restricted space and unable to expand. Once you crossed the border, however, you entered a new territory—a seamless space that could expand indefinitely along the coast through the construction of new piers—but also inland through the development of a new freight centre and a new commercial train line. Significantly, although still formally in Greece, in this new territory the normal national laws and provisions applying to production and labour were suspended. 'A territory within a territory,' in which containers were only in transit. This invisible dividing line has become for several years the site of intense labour struggles that reproduced preexisting masculinist and nationalist labour union subjectivities but also opened up possibilities for new political borderline subjectivities of cybernetic labour.

Although workers in OLP faced considerable salary reductions as a result of austerity policies implemented since 2009 in Greece, they continued to enjoy indefinite contracts of service, well-paid salaries, full social security rights, paid overtime, and the right to unionize and strike. Labour rights granted to OLP workers were the fruits of decades of labour struggles. The protagonists of these struggles were robust Greek males able to employ tactics to control the limit between normal working hours and overtime. There were very few women—usually contained in the cruise sector and in administrative services—and even fewer migrants, the latter almost exclusively employed by cleaning agencies holding subcontracting agreements with OLP. Labour in OLP presupposed a masculine subject devoid of domestic and care responsibilities but also equipped with the national citizenship and labour rights granted and guaranteed by a sovereign state.

The new labour regime established after the arrival of COSCO in Piraeus involved a complex web of contracting and subcontracting labour agreements. Other than a few full-time employees, most of the dockworkers are recruited and trained by the logistics company *Diakinisis* (in its 2014 annual financial statement, PCT declares that it employs only 261 workers while the total number of people employed in the PCT terminal is estimated to be around 900–1000). Workers in COSCO are precarious in the sense that they have short-term, limited-time or temporary contracts and are informed about their working schedule each week according to logistical requirements (Parsanoglou 2014). Precarity in the port is, however, produced within a fixed and hierarchically regulated time-space. It differs significantly from the precarity of workers in the creative and digital industries and has more affinities with the work of cleaners in subcontracting companies in Greece. Although work and leisure time are intertwined, they are fixed and separated in ways that prevent overtime and extra payment to workers. The model of time management is not that of digital industries, where the time-space of work invades that of private and leisure time. Precarious workers in Piraeus have specific schedules and are expected to be docile in following predetermined protocols and routines. They are unable to control the conditions of time and space in the workplace in order to maximize their earnings and demand labour rights as the labour unions in Piraeus did before the COSCO concession.

OLP's masculine subjectivities are aiming at reterritorialization that is nurtured on a desire to annul the concession and renationalize the port. This desire tends to be also blurred with anxieties over both the prevailing narratives of 'precarity as Chinification' and encroaching feminization. Against these threatening prospects, the unions have organized large mobilizations that have delayed the implementation of the concession. Alongside the 'Chinese Go Home' graffiti, feminized and racialized images are posted on the office walls of OLP dockworkers as iconic commentaries on the broken relationship between the sovereign state and labour unions. The predominant regime of nation, gender, and class has collapsed, leaving in its wake a set of impossible struggles. While the port's long tradition of masculinized labour with full labour rights has been left to mock and mourn the advent of 'feminized' labour on the other side of the invisible border, it appears also to have been stripped of its right to unionize and/or ability to coerce the sovereign government to concede to its industrial demands.

By bringing to our attention this invisible dividing line between the two sides, the female trade unionist who guided us also tried to explicate the porousness of the border. The OLP union's official discourse is aimed at a reterritorialization, that is, the reestablishment of a well-defined limit between stable, masculine full-time employment and the emerging precarious forms of feminized labour. Yet, she insisted on elaborating on the ways in which the border is being crossed and defied. COSCO's piers, as she explained, were unbounded, and were constantly expanding in different directions, and in time, by introducing new connectivities, technologies, and forms of control. OLP, on the contrary, was suffocating economically under the concession because of its limited one-pier space and its stubborn insistence on continuing to impose strict limits between 'normal' full-time and precarious forms of employment.

Since our research visits, the time-space border between the OLP and the COSCO piers has faded, under pressure of a new machinic assemblage emerging in the port, and so have the strict gendered and racialized divisions of labour. Logistical operations are undermining the dominant presupposition of the port as a space where physical (masculine) strength is required. As the entanglement of machines, people, and things accelerates, the labour struggles—aiming at the reterritorialization of the port and the reestablishment of the border—have been increasingly unsuccessful. As Haraway argues, 'the success of the attack on relatively privileged, mostly white, men's unionized jobs is tied to the power of the new communications technologies to integrate and control labour despite extensive dispersion and decentralization' (Haraway 1991, 305).

In other digitized industries, single women have replaced men because they are considered as to be ideal docile workers. Feminization is also about transforming male labour into a complex cybernetic organism that shares the stereotypical docility of female workers. Unlike the cyborgs—the utopian creatures of a post-gender world so often celebrated in gender theory—Piraeus' cyborgs are precarious, masculine, and docile. They obey protocols and not rules. They are not working on an assembly line, although they perform repetitive tasks made possible through the digitization of all their functions. The economic crisis in Greece accelerates this process of cybernetic transformation of docile workers: recession and austerity have hit predominantly younger generations and male-dominated sectors—such as the construction and ship building industries—and have created a vast pool of unemployed young males willing to work without the labour rights of previous generations of male workers. As the port is transformed into a transit place of increased productivity and machinic efficiency, new forms of labour vulnerability, that were previously reserved only for women and migrants, become part of the everyday reality of Greek males too.

The arrival of COSCO-PCT in the port does not only create desires for reterritorialization, but it also animates futuristic scenarios of the transcendence of the current crisis (Grappi 2015). The desire to work in the port continues in the accounts of male residents of the Perama hills overlooking the port of Piraeus, an area hit by recession, austerity, unemployment, and poverty following the collapse of the ship repair industry (Hatzopoulos and Kambouri 2014). Male residents remember with awe how the enormous cranes arrived in pieces at Pier II. In just a few days, they were unloaded from their ships, assembled, and made functional, almost without human intervention. Observing the mechanical transformation of the port from afar fuels expectations and desires that new jobs, new roads, and new shopping centres will be created in the area.

In futuristic scenarios, the COSCO concession is expected to bring about spectacular technological transformations, and Piraeus is represented in these narratives as an unbounded and deterritorialized space, where modern machines and skilled humans will work in harmony, optimize production, and maximize efficiency. Such scenarios project desires of a seamless space, where logistical processes will enable the complete control of labour from a distance and unlimited technological progress across time. The COSCO concession becomes, thus, a vehicle that will relocate Piraeus outside the Greek territory and the economic dead-end of the Greek sovereign debt crisis. The theme of the 'territory within a territory' resurfaces, but this time as a neoliberal utopia of seamless optimization.

The expectations created by such futuristic visions are supported by references to 'greater-than-expected' growths in traffic and economic efficiencies. Data presented by the Commercial Director of PCT showed that these changes in the port brought about a dramatic increase in productivity, especially in Piers II and III (Vamvakidis 2013). Piraeus ranked in 2014 as the 39th busiest port in the world and the third busiest container port in the Mediterranean (after Valencia and Algeciras in Spain) and Port Said. The port posted a 13 per cent increase in container traffic in 2014, on top of a 20 per cent increase in 2013, and a massive 77 per cent rise recorded in 2012 (the highest in the world).

Technological transformations require that Piraeus should grow into a transit place unbounded by national and European borders, interests, and restrictions. As Greek imports and exports continue to shrink, growth becomes synonymous with growth in transhipment operations. Containers entering the port of Piraeus do not usually enter the Greek territory (Vamvakidis 2014). The location of the port is considered ideal for the creation of this transit place of unlimited potential since, both geographically and politically, Piraeus is viewed as a major 'gateway' to Europe. As Captain Fu Cheng Qiu, the CEO of PCT, states, 'No other country in Europe offers such potential. We believe that Piraeus can be the biggest port in the Mediterranean and one of the most important distribution centres, because it is the gateway to the Balkans and southern Europe' (quoted in Smith 2014). The concept of the 'gateway to Europe' is double-edged as it refers both to the strategic geographical position of Piraeus and to the precarious political condition of a state in crisis that has been released from the pressure to respect established labour rights.

Futuristic, deterritorialized visions of the future make possible plans about the expansion and dispersion of technological advancements outside the limits of the port. The COSCO concession stimulates further interest for Chinese investments in infrastructure that may be given up to privatization in the context of austerity and structural readjustment policies. A new freight centre—currently completed but unused—in the Thriassio plain is linked to the port through a new rail link constructed in 2014 with government and EU funds. According to PCT management and the Greek government, such infrastructure opens up unlimited new possibilities for the transport of goods via train and truck to Central and Eastern Europe. The continuing expansion of such infrastructure within and outside the container port will probably make Piraeus the busiest port in the Mediterranean in the coming years.

The Piraeus Container Terminal as a Machinic Assemblage

The 'Industry and Idleness' engravings created by William Hogarth in 1747 narrate the life stories of two characters, Francis Goodchild and Thomas Idle, in a series of 12 plates (Tate Gallery 2016). In the first plate, entitled 'The Fellow 'Prentices at their Looms,' the innate qualities of the two characters are juxtaposed. Set in a small room of a textile workshop, presumably in Spitafields, London, this engraving depicts the industrious Francis working energetically over a weaving loom while Thomas has fallen asleep standing up in front of his idle loom. Standing behind the men and their looms, their boss, the master weaver, is holding a fat wooden stick and looking disenchanted at the image of stillness of Thomas and his loom. In contrast to the dynamic image of Francis and loom harmoniously weaving what seems to be an intricate fabric, Hogarth offers an excessive image of non-work, signified by a cat who is playing with the shuttle of the loom, a clay pipe that is wedged into the handle of the loom that is blocking the possibility of it being set to motion, and a mug of ale—presumably the cause of Thomas' sleepiness—that rests on top of the loom.

Let us depart from the intended rationale of Hogarth's linear tale that ends up with the glorification of industry, with Francis becoming the Lord-Mayor of London, and the denunciation of idleness, Thomas being executed by hanging. We would like to use Hogarth's tale as a foil to discuss what we deem to be critical to the analysis of precarious labour in regimes of logistical governance such as those that are being shaped in the port of Piraeus.

Hogarth's representation of the weaving room is ostensibly humanistic. The efficiency of the production process for manufacturing fabric is primarily linked to human activity: industry or idleness. The control of labour is intrinsically attributed to the presence of the master weaver inside the working place, his gaze constitutes the logic of surveillance over the activities of the workers, while his very presence embodies the disciplinary power he can exert if the workers fail to behave according to the standard (presumably by use of the wooden stick). The motionless loom, the wooden stick as a weapon-machinery that threatens bodily punishment, or even the cat who attempts to set the loom in motion by playing with its wedge, are ultimately treated by Hogarth as aesthetic appendages for adding elements of irony and mockery to his moral, humanistic tale. Our argument, in this section, is that looms, wooden sticks, and even cats are constitutive of labour processes and relations. Tracing the multiple becomings of looms, wooden sticks, and cats within production processes can reveal the complex entanglements that also generate the subjectivity of labour within contemporary, globalized container ports.

The organization of production in the container terminals of Piraeus port no longer rests on human labour utilizing machines to perform tasks. Of all the human workers active within the space of the container terminal, only stevedores executing the task of locking and unlocking containers before they are moved in and out of ships perform proper manual work (Gogos 2014). All other activities—including the unloading, loading, and stacking of containers, the mobilities of different types of machines such as gantry and stacking cranes, and the operation of transport vehicles—are planned and operated by so-called Terminal Operating Systems (TOS). These TOS, which profess to offer an all-encompassing handling of the planning, operational, and monitoring work performed in the terminals, are based on 'smart' algorithms for modelling, modifying, and solving optimization problems (like IBM's ILOG suite utilized by NAVIS).

Two software operating systems are employed in Piraeus port to command and control all movements of machinery, things, and people within the space of the container terminal. They are CATOS, which is the TOS platform used by PCT (operative since 2009) and developed by Korean-based maritime logistics solution company Total Soft Bank Ltd (TSB), and NAVIS Sparks N4 which is used by OLP (operative since 2001) and was developed by the US company NAVIS. They constitute, as Soenke Zehle and Ned Rossiter would put it, an algorithmic apparatus 'designed to extract value, orchestrate efficiencies, and optimize productivity' within the logistical site of the Piraeus container terminal (Zehle and Rossiter 2015).

Along these lines, we can consider operations in the Piraeus container terminals as a social machine, or a machinic assemblage in the line of Deleuze and Guattari's discussion of Kafka's minor literature (1986). The container terminal takes human labour in its gears, along with containers and all the machinery that moves them around the terminal: quay cranes, rail-mounted and rubber-tyred gantry cranes, trucks, and software platforms that generate and control these complex movements through algorithmic computations.

The deployment of the concept of machinic assemblage diffuses the centrality of human labour in the production process and highlights the limitations of imagining social struggle through a human-centric approach. As Maurizio Lazzarato succinctly puts it, 'the humanization of work ... has nothing progressive about it' (2014, 120). Within the optimization regime of production in Piraeus, it is indeed 'not clear who makes and who is made in the relationship between human [software] and machine. It is not clear what is mind and what body in machines that resolve into coding practices' (Haraway 1991, 313).

Anna Tsing's concept of the figurations of labour can be useful here for breaking our analysis further out of a humanistic horizon. Tsing's figurations of labour can alternatively reveal how contemporary strategies for the control of labour 'involve nonwork tropes, particularly tropes of management, consumption, and entrepreneurship, becoming key features in shaping informatised labor in logistical sites' (Tsing 2009, 151).

Tracing the figurations of labour that are produced within the regimes of control in the Piraeus container terminals, one such figure that emerges is that of the 'remote operator.' The 'remote operator' is typically a machine operator—of a crane, a truck, a straddle carrier. Their role is that of sustaining the spectacle of the informatized, machine-enabled efficiency of the container terminal. Their body thus occupies semi-visible positions with a bird's eye view of the container terminal; confined within small cubicles appended to container handling machines, their encapsulated body ensures that the machine operates in an optimal fashion.

Another figuration of labour emerging in the Piraeus container terminals is that of the 'gamer-programmer.' The 'gamer-programmer' usually inhabits a desk in the operations department of the terminal operator, where they are more persistently exposed to the graphical user interface (GUI) of the TOS. They might be involved in vessel planning or yard planning and so on. They tend to attribute the emergence of contingencies that cannot be handled by the software system to human errors, usually made by 'remote operators.' The 'remote operator' can, however, easily step into the figure of the 'game-programmer,' assisted by virtual reality training simulators and intelligent remote control stations (as the 'Crane Simulator' developed by Total Soft Bank for training crane operators) (Total Soft Bank 2016a).

Following Deleuze and Guatarri's reading of Kafka, we should see these figurations of labour as part of the machinic assemblage not 'only in their work but even more in their adjacent activities, in their leisures, in their loves, in their protestations, in their indignations, and so on' (1986, 81). Protest, or the emergence of social struggle, is enunciated within the machinic assemblage. Similar to the woman's question to K. in Kafka's *The Trial* (1986, 82), 'Is it reforms you want to introduce?' the possibility of the dockworkers' resistance to the control of labour in the container terminals forms part of an assemblage composed also of their revolting activities, of machines, and of their desire for liberation.

In the machinic assemblage of the Piraeus port, the 'remote operator' becomes the protagonist in a very telling episode of collective rebellion inside the terminal (the first recorded collective rebellion against the labour regime enforced by PCT in response to the prohibition on unionization and collective bargaining and the use of subcontracted labour). Rebellion is sparked by a machine failure, albeit the failure of an inferior machine. Some PCT workers decided to form a five-member workers' committee to complain about working conditions. These workers, mainly rail-mounted gantry crane operators, were prompted to act because of the breakdown in the heating and air-conditioning system in the operator's cabin. The broken machinic supplement was the straw that broke the camel's back since, as they claim, their sensitive job depended on 'optimal temperature conditions' (Batsoulis 2014).

Although this episode can be inscribed in the framework of traditional labour struggles for better working conditions, it simultaneously exceeds this interpretation. The demand for the fixing of the heating and air-conditioning in the operator's cabin is simultaneously inscribed in the discourse of the optimization of terminal performance, where the workers' claim takes optimization to its limits and demands optimization everywhere.

MACHINIC IDLENESS

The spectre that haunts the organization of work at the Piraeus container terminal is machinic idleness. Powerful algorithms and data mining techniques continuously document and analyze machinic behaviour with the aim of minimizing the time spent in idleness due to the entanglements of machines, people, things, and software.

In Total Soft Bank's jargon (the IT company that developed the TOS used by PCT in Piraeus), machinic idleness is articulated as the 'empty travelling time of equipment' or, on other occasions, as the state of a machinery being 'idle' (i.e., motionless) within the space of the terminal. The company's new optimization software, CHESS—Container Handling Equipment Supervisor System, is its proposed solution to minimizing idleness by precisely increasing performance and efficiency in semi- or fully automated container terminals (Total Soft Bank 2014). Against this declared scope of CHESS' application—supervising equipment—we have to underline that its approach towards minimizing idleness always targets amalgams of software, machines, and workers. The idleness of a quay crane, as conceived and calculated by CHESS, involves concomitantly the possible idleness of a functional whole that includes the quay crane, a possible glitch on the software that controls the crane, the activities of the crane's human operator, an RFID reader and tag placed on the quay crane, and so on. A concrete example of this extended scope of application of CHESS comes through its analysis in the operations of the Guangzhou South China Oceangate Container Terminal, where one of the conclusions reached by the software system was that the idleness of yard trucks was primarily caused by a particular driver's negligence (Total Soft Bank 2016b).

Along these lines, idleness becomes critical in the optimization regime of Piraeus port, primarily in its machinic guise, and cannot be distinguished as, merely, the non-work of labourers or as the non-work of mechanical equipment. In this respect, machinic idleness constitutes the central concern for the control of labour, while the amalgam of machines, people, and software is instituted as the subjectivity of labour.

Our central argument in this section is that the analysis of the control of labour in contemporary global supply chains needs to be rethought. We propose that, first this analysis should recognize machinic assemblages as the principal plateau of control practices. Second, that its focus should move beyond an understanding of the informatization of labour, based on a uniform view of social networks as distributed systems, which emanates from the (often thought as ubiquitous) model of the Internet. And third, that the control of labour takes the form of the control of machinic idleness.

This set of arguments diverges from a series of recent analyses that address, directly or indirectly, the forms and modes of the control of labour in contemporary capitalism. Take, for example, the humanistic analysis of Carlo Vercellone (2010) in which the machinic elements of labour processes are ultimately devalued. The control of labour under 'cognitive capitalism' is deemed by Vercellone to be exercised through the process of the financialization of everyday life and by instituting precarity as a structural element of the contemporary neoliberal regime. It involves, ultimately, pure and simple coercion (106). Vercellone goes so far as to argue that the predominance of this logic of financialization leads to increasing autonomy in the organization of production and business firms being concerned primarily with their financial architectures and the optimization of the allocation of their financial resources (which is realized via algorithmic operations of financial trading, one may add).

Tiziana Terranova's (2014) study on the relationship between algorithms and capital, on the other hand, recognizes the critical importance of linking together technology and processes of subjectivation. Terranova's inquiry into how contemporary algorithmic automations are constituted through modes of control and monetization, falls, however, ultimately into (a) a discussion of the architecture of the Internet as a model for the emergence of new, heterogeneous forms of sovereignty, of Internet protocols such as TCP/IP as critical technologies in these emerging forms of organization, and of proprietary social networks as key drivers of digital socialities, and (b) an understanding of political action that embraces the 'enactment of new subjectivities' for the democratic reappropriation of technologies and institutions (Terranova 2014).

A clear break with these approaches is accomplished by Maurizio Lazzarato's (2014) recent work, particularly the introduction of the concept of 'machinic subjectivity,' whose 'operations can only be determined by way of the functional whole of humans-machines' (99). By fully endorsing the notion of 'machinic subjectivity' as exemplary of contemporary labour subjectivity, we can insist on the non-reducibility of the non-human agents and the non-human components of machinic assemblages and prevent these from being subsumed under the horizon of humanized analyses and politics. To put it differently, the approaches of Vercellone, Terranova and others are, too often, fixed on studying processes of subjectivation in connection to processes of production of the amalgams of humans, machines, and software. These analyses are, thus, neglecting not only the role of processes of desubjectivation in shaping the contemporary control of labour but also the ways in which desubjectivation can generate strategies for resisting or eluding control. Lazzarato urges us, along these lines, to 'seize the opportunity of desubjectivation opened by machinic enslavement so as not to fall back on the mythical-conceptual narratives of producers, workers, and employees' (2014, 114).

Our study, on the intersections of logistical technologies and labour regimes in Piraeus' container terminal, allows for a slightly different perspective in relation to Lazzarato's privileging of the financial trader's subjectivity as somehow emblematic of contemporary capitalism. We would like to inquire, instead, into the forms of machinic subjectivity when this 'machinic' element is not universally associated with the distributed topologies and networks, the flexibility and openness, that is often taken to be the invariable condition of networked machines (as if a certain view on the architecture of the Internet is a universal given). Lazzarato mentions 'mathematical systems, data banks, interconnected computer networks, telephone networks, and so on' as the non-human components of the trader's machinic subjectivity (2014, 125). But, what if we study machinic subjectivities whose non-human components are more complex in terms of the relationship and communication amongst their different elements? What if, in other words, we give more emphasis to the specificities of the functioning of different types of machinic subjectivities?

The machinic subjectivities emerging from the entanglements of machines, people, things, and software in Piraeus port present a productive starting point for the analysis of contemporary mechanisms of the control of labour. The crucial question to address, here, is how the control of labour is exercised as the control of machinic idleness. To do this we first need to critically reflect on the notion of protocological systems as synonymous to digital networks in general, or to protocols as the central forms of control of digital networks (Galloway 2004; Galloway and Thacker 2007). Our claim is that protocological control is only one amongst the differing organizational logics that produce the complex software regimes that are in place in the Piraeus terminals.

On the one hand, protocols are central in controlling all information and data exchanges that organize the real-time operation of the terminal. In the semi-automated environments of the Piraeus container terminals, there are several types of protocols that are instituted to govern the data flows within the TOS and to ensure interoperability with other software systems (Total Soft Bank 2016c). Radio link protocols enable data exchange between handling equipment, human agents, and the TOS within the terminal. Electronic Data Interchange (EDI) protocols enable the interfacing of the TOS with information and transactions about containers, routes, quality controls, customs procedures, and so on that involve the entire user community of the port. In spite of their centrality in managing these data exchanges, protocols are not critical as the organizing principle of the software-driven promise of the continuous boosting productivity and the perpetual optimization of container terminal operations. These processes are built into the software architectures of TOS and their management relies much more on the control of machinic behaviour rather than on the regulation of data flows within the network as a whole.

We have to note, here, that the digital networks in the terminals have different topologies compared to other types of digital networks—let's say the Internet. CATOS is designed, for instance, as a decentralized system, with a few central nodes and with access limited to authorized users only. These central nodes include a radio server, a database server, a dispatcher server, and a communication tower channelling radio messages. Apart from their privileged position within the CATOS network's architecture, these nodes play a critical role in the processes that spur the promise for the continuous reconfiguration of the mobilities of things, machines, and people within the terminal. These nodes have also a storage function for all operational data documented by the TOS, which are then mined for producing performance analyses and subsequent proposals for the optimization of the terminals' operations.

The predominance of a logic of hierarchizations that governs the entanglements of software, machines, things, and people surfaces also in the discourse of efficiency and optimization. In an excerpt from a NAVIS report entitled 'The Road to Optimization,' a particular type of rigid hierarchization becomes evident: 'For most terminals the key to achieving higher productivity is not to operate equipment faster, but to make the operation more consistent, with fewer delays while equipment is waiting for containers or other equipment' (NAVIS 2014). The minimization of equipment waiting time is thus prioritized over the acceleration of equipment operation. In this sense, it is equipment that comes first, with (as explained in another optimization report) quay cranes coming before other types of equipment as the most expensive and the least flexible machinery.

The control of precarious labour in the port of Piraeus cannot be thus associated with notions of flexibility, distributed networks, and incessant data flows. Contrary to this imaginary, forms of control in Piraeus are based on a combination of flexible and rigid hierarchizations, distributed and decentralized topologies, and mining techniques and projections on machinic behaviour.

Conclusion

On 18 July 2014, workers at the PCT container terminal organized their first strike. Their actions addressed both COSCO and *Diakinisis*, their formal employer, acknowledging the different layers of control in the port but also their interconnections. The next day, COSCO-PCT promised to meet some of their demands and negotiate others. As a result, the strike ended. One of the achievements of this action was the founding of the first trade union representing at least some workers in the PCT (ENEDEP 2016). The union continues to fight, with its latest action being the (averted) call for a walkout in Piers II and III in February 2016, but it has still not managed to force PCT to sign a collective labour agreement, impose a training process, or achieve any major improvement of health and safety conditions (Frantzeskaki 2016).

As the concession expands in time and space—to absorb the cruise, car and passenger terminals, the ship repair industry, the train line, the Thriassio freight centre, and commercial buildings in the surrounding area—the machinic assemblage becomes more and more intense in integrating multiple political subjects, labour histories, and tropes into this precarious new labour regime. This complex process can no longer be addressed simply as a question of national sovereignty or the upholding of the unionism of Greek male dockworkers. Nor can it be explained solely as a form of Chinese economic neoimperialism that imposes a single labour regime across continents. What is clear is that despite COSCO's impact on the digitization and transformation of labour relations in the port, the specific forms that precarity takes is not independent from Greek labour struggles and logistical operations.

At this moment of the intensification of the precarious labour regime pervading Piraeus, it becomes urgent to focus on the new machinic forms of control of labour that organize the port and to reimagine new political subjectivities that may emerge. What we have tried to show is that in order to grasp the transformations in Piraeus, we may be required to distance ourselves from humanistic thinking, and territorial conceptions of space, and begin to engage with the details of the precarious lives of cybernetic organisms, gender crossings, and processes of deterritorialization and reterritorialization. Thinking of Piraeus as a machinic assemblage forces us to rethink the one-sided conception of precarization as a negative process and makes possible a reimagining of precarious workers as cyborgs. The notion of a cyborg is in turn double-edged: neither innocent nor evil, it can be a docile unit, while also becoming part of political struggles that move beyond the confines of nationalist, masculinist and racialized conceptions of labour. Thinking of the port as a machinic assemblage twists the neoliberal futuristic utopias, turning them into nightmares of precarization, but at the same time giving rise to new political processes of becoming, out of which new kinds of struggles may be born.

References

Batsoulis, Dimitris. 2014. (Fired ex-PCT employee in discussion with authors) 16 February 2014.
Burgoon, Brian,and Damian Raess. 2014. 'Chinese Investment and European Labor: Should and Do Workers Fear Chinese FDI?' *Asia Europe Journal* 12, nos. 1–2: 179–97.
Child, John. 2009. 'Context, Comparison and Methodology in Chinese Management Research.' *Management and Organization Review* 5, no. 1: 57–73.
Deleuze, Gilles and Félix Guattari. 1986. *Kafka: Toward a Minor Literature.* Translated by Dana Polan. Minneapolis: University of Minnesota Press.
Deleuze, Gilles and Félix Guattari. 1999. *A Thousand Plateaus: Capitalism and Schizophrenia.* Translated by Brian Massumi. Minneapolis: University of Minnesota Press.
ENEDEP (Union of Dock Workers of Piraeus). 2016. Accessed 28 June 2016. http://enedep2014.blogspot.gr/
Frantzeskaki, Anastasia. 2016. 'The Privatization of the Piraeus Port Authority in Greece: What's Really Happening.' *Tlaxcala.* Accessed 1 January 2017. http://www.tlaxcala-int.org/article.asp?reference=17002
Galloway, Alexander R. 2004. *Protocol: How Control Exists After Decentralization.* Cambridge, MA: MIT Press.
Galloway, Alexander R., and Eugene Thacker. 2007. *The Exploit: A Theory of Networks.* Minneapolis: University of Minnesota Press.
Gogos, Giorgos. 2014. (General Secretary of OLP dockworkers Union), in discussion with authors, 12 February 2014.

Grappi, Giorgio. 2015. 'Logistics, Infrastructures and Governance after Piraeus: Notes on Logistical Worlds.' *Logistical Worlds: Infrastructure, Software, Labour.* Accessed 28 June 2016. http://logisticalworlds.org/blogs/logistics-infrastructures-and-governance-after-piraeus-notes-on-logistical-worlds

Haraway, Donna. 1991. 'A Cyborg Manifesto: Science, Technology and Socialist-Feminism in the late 20th Century.' In *The Cybercultures Reader*, edited by Barbara M. Kennedy and David Bell, 291–323. London and New York: Routledge.

Hatzopoulos, Pavlos, and Nelli Kambouri. 2014. 'The Logistical City from Above.' *Logistical Worlds: Infrastructure, Software, Labour.* Accessed 28 June 2016. http://logisticalworlds.org/?p=216

Meunier, Sophie. 2012. *Political Impact of Chinese Foreign Direct Investment in the European Union on Transatlantic Relations.* Brussels: European Parliament.

NAVIS. 2014. 'The Road to Optimization.' Accessed 28 June 2016. http://navis.com/news/in-news/road-optimization

Neilson, Brett. 2018. 'Precarious in Piraeus: On the Making of Labor Insecurity in a Port Concession.' *Globalizations*: forthcoming.

Lazzarato, Maurizio. 2014. *Signs and Machines: Capitalism and the Production of Subjectivity.* Translated by Joshua David Jordan. Los Angeles: Semiotext(e).

Parsanoglou, Dimitris. 2014. 'Trojan Horses, Black Holes and the Impossibility of Labour Struggles.' *Logistical Worlds: Infrastructure, Software, Labour.* Accessed 28 June 2016. http://logisticalworlds.org/wp-content/uploads/2015/04/535_UWS_Logistical-Worlds-digest-2014-v10-WEB.pdf

Piraeus Container Terminal SA – PCT. 2016. 'Corporate Philosophy.' Accessed 28 June. http://www.pct.com.gr/content.php?id=7

Smith, Chris, and Yu Zheng. 2016. 'Chinese MNCs' Globalization, Work and Employment: with Special Reference to Europe.' In *Flexible Workforces and Low Profit Margins: Electronics Assembly between Europe and China*, edited by Jan Drahokoupil, Rutvica Andrijasevic, and Devi Sacchetto, 67–92. Brussels: The European Trade Union Institute.

Smith, Helena. 2014. 'Chinese Carrier COSCO is Transforming Piraeus – and has Eyes on Thessaloniki.' *The Guardian*, 19 June 2014. Accessed 28 June 2016. https://www.theguardian.com/world/2014/jun/19/china-piraeus-greece-cosco-thessaloniki-railways

Tate Gallery. 2016. 'Hogarth: Hogarth's Modern Moral Series, Industry and Idleness.' Accessed 28 June. http://www.tate.org.uk/whats-on/tate-britain/exhibition/hogarth/hogarth-hogarths-modern-moral-series/hogarth-hogarths-3

Tsing, Anna. 2009. 'Supply Chains and the Human Condition.' *Rethinking Marxism* 21, no. 2: 148–76.

Terranova, Tiziana. 2014. 'Red Stack Attack! Algorithms, Capital, and the Automation of the Common.' *Quaderni di San Precario*, 18 February 2014. Accessed 28 June 2016. http://www.euronomade.info/?p=2268

Total Soft Bank. 2014. 'CHESS Makes its First Move from GOCT.' Accessed 28 June 2016. http://www.tsb.co.kr/RBS/Fn/CommBoard/Download.php?-RBIdx=Ver1_50&Idx=14&FIdx=31

Total Soft Bank. 2016a. 'Overhead Crane Training Based on Virtual Reality!' Accessed 28 June 2016. http://www.tsb.co.kr/RBS/Fn/FreeForm/View.php?RBIdx=Ver1_61

Total Soft Bank. 2016b. 'Sail to the Future.' Accessed 28 June 2016. http://www.tsb.co.kr/RBS/Data/Files/fnAAN/News05/Final_20130402.pdf

Total Soft Bank. 2016c. 'CATOS.' Accessed 28 June 2016. http://www.tsb.co.kr/Ver1/Products/0101-2.php

Vamvakidis, Tasos. 2013. 'Piraeus Container Terminal S.A.: The South East Gate of Europe.' Presentation at the European Maritime Week Conference, Athens, Greece, 20–24 May 2013. Accessed 28 June 2016. https://www.slideshare.net/jmceunipi/tassos-vamvakidispiraeus-container-terminal-sa-the-south-east-gate-of-europe

Vamvakidis, Tasos. 2014. (PCT commercial director), in discussion with authors, January 2014.

Vercellone, Carlo. 2010. 'The Crisis of the Law of Value and the Becoming-Rent of Profit.' In *Crisis in the Global Economy: Financial Markets, Social Struggles, and New Political Scenarios*, edited by Andrea Fumagalli and Sandro Mezzadra. Translated by Jason Francis Mc Gimsey, 85–118. New York: Semiotexte.

Zehle, Sonke, and Ned Rossiter. 2015. 'Mediations of Labor: Algorithmic Architectures, Logistical Media, and the Rise of Black Box Politics.' In *The Routledge Companion to Labor and Media*, edited by Richard Maxwell, 40–50. New York: Routledge.

Zhu, Cherrie Jiuhua, Mingqiong Zhang, and Jie Shen. 2012. 'Paternalistic and Transactional HRM: The Nature and Transformation of HRM in Contemporary China.' *The International Journal of Human Resource Management* 23, no. 19: 3964–82.

CHAPTER 9

Asia's Era of Infrastructure and the Politics of Corridors: Decoding the Language of Logistical Governance

Giorgio Grappi

This story begins in Baku, the capital of Azerbaijan on the shores of the Caspian Sea where, after the construction of the 'flame towers' in 2012, a large set of new skyscrapers is expected to transform the urban skyline (Skyscraper City 2016). They represent a new kind of development that is taking Azerbaijan by storm. In fact, the constant flow of financial activity pouring billions of dollars into the country is giving Azerbaijan a key role inside a new Asiatic geo-economy. The new forms that are casting their shadows on the Caspian Sea, and the renderings made available by promotion videos, are indeed the 'markers' of the appearance in Baku of what Keller Easterling has described as 'the software of Extrastatecraft' or a 'transshipment landscape' (Easterling 2014, 46). More than simple buildings, they send a message to the global business audience that the country is now seriously committing to its role within 'global assemblages' of power (Sassen 2006). The country is also taking other important steps to

---

G. Grappi (✉)
Department of Political and Social Sciences, University of Bologna, Bologna, Italy

© The Author(s) 2018
B. Neilson et al. (eds.), *Logistical Asia*,
https://doi.org/10.1007/978-981-10-8333-4_9

demonstrate its availability to be included in transnational infrastructural projects that focus on Central Asia as a new crossing point. As recently reported by the global consulting agency Information Handling Services (IHS), the projected Baku International Sea and Trade Port is expected to offer 'opportunities in China-Europe overland links' (Mooney 2015). The port, the Azerbaijan government says, expects to benefit from the overland links between China, Central Asia, and Europe related to Chinese initiatives to build a New Silk Road across Asia and will be built following the port-centric logistics corridor model promoted by DP World at Jebel Ali port in Dubai (Sulayem 2013). This model is implemented as a logistical manufacturing centre where all activities of finishing, processing, and packaging will be realized. The area will include a special economic zone (SEZ) and is designed 'as a hub where companies use locally produced materials to add value to imported products before shipping them onwards to destinations in Central Asia, Europe, and Turkey' (Mooney 2015).

Taleh Ziyadov, Director-General of Baku International, highlights that the whole project is clearly connected with the New Silk Road discourse and its position in the global logistical chain: 'an important point about the silk road revival,' says Ziyadov, 'is that it works on the basis that all countries improve their infrastructure and want to harmonize policies across trade, customs, and transit procedures. If Georgia doesn't have good ports and roads, then we cannot reach our potential. Likewise, if the links between Kazakhstan and China aren't good enough, we will not benefit' (Mooney 2015). This suggests that one hundred years after 1900 delegates from across Asia and Europe met in Baku for the Congress of the Peoples of the East in 1920 and committed to support anti-colonial movements, the city is finally becoming an intersection in the restructuring of Asian political space. In fact, Baku is at the crossroads of different infrastructural projects, promoted by a variety of state actors and international institutions, that are converging around the idea that logistical corridors connecting Eurasia will bring a new era of development for the region and the world.

Despite their very different political and institutional genesis, projects such as the 'Belt and Road Initiative' (BRI) or 'One Belt, One Road' (OBOR)—the Chinese official name of the New Silk Road, the Central Asia Regional Economic Cooperation Corridor (CAREC)—initiated by the ADB, and the Transport Corridor Europe-Caucasus-Asia (TRACECA)—a side project of the European Union Trans-European

Networks (TEN) reveal how 'Asia's Era of Infrastructure' (to quote the title of a panel organized during the World Economic Forum held in Davos in January 2016) is more than a simple proposition of regional development. These projects, the way they are proceeding and the imaginary they mobilize, share an inner logic that views existing institutions, their territorial dimensions, and the temporality of institutional politics as embedded in the infrastructural and operational dimensions of logistics. In fact, if their economic justification is to cut transit times from China to Europe, and bring new economic activity along the way, what they show is an emerging reorganization of global political spaces that I propose to term 'the politics of corridors.' To observe the coherence of this reorganization is not to argue that a single model is being imposed across the globe. On the contrary, I argue that, against the background of a discourse that pretends to generate smooth and homogeneous dynamics, the 'politics of corridors' encounters obstacles, resistances, and frictions that result in variations of form and operation. Moreover, while such corridors are emerging globally, we must observe that while different corridor initiatives meet, overlap, conflict, and diverge, they are producing a variegated geography of logistical power. Nevertheless, with this proposition I aim to stress how a new political formation is emerging within and around these corridors. And furthermore, that this formation can be considered as a global logistical institution that constantly reckons with the state form, forcing it 'to negotiate their role with a multifarious array of agencies and reckon with heterogeneous legal orders, logistical protocols, financial algorithms, and monetary arrangements that exceed the control of any state' (Mezzadra and Neilson 2014).

If the geopolitical scenario suggests competition between China and other regional players regarding the control of supply corridors across Asia, I argue that we must pay attention to the different logics at work in the *background* and that, through this analysis of global dynamics, the very concept of 'Asia' as a political and geographical unity begins to blur. Logistical operations along global supply chains have reshaped the processes of production. They are also contributing to the formation of a new type of hybrid political discourse, which operates behind the technicalities related to the promotion of intermodal solutions and the removal of *bottlenecks*. This discourse finds a synthesis around the image of 'corridors' that is giving rise to new political forms that are transforming relations between private economic activity, state intervention, and international institutions. While the 'supply chain' refers to the production processes that exist behind the pervasive presence of logistics, the language of 'corridors' refers to the materiality of infrastructure and the so-called soft infrastructure of governance that makes logistical operations possible at a larger scale. As we will see, while different actors with different geopolitical and economic interests are at play, a new consensus presents corridors and their operations as structural priorities.

Corridors, Supply Chains, and Logistical Governance

The language and image of corridors is an emerging catchall in the discourses of logistics. From trade and investment corridors to freight corridors, from digital corridors to development corridors, from transport corridors to industrial corridors, it is difficult to avoid reference to this concept. As Deborah Cowen (2014) observes, 'logistics corridor projects and their visual rendering in technical and popular cartography are popping up all over the world.' These maps, Cowen suggests, 'craft a different spatial imaginary than blocks of transnational territory' and the image of corridors as physical infrastructures is insufficient to grasp their meaning. In fact, material intervention proceeds jointly with a relevant and increasingly sustained focus 'on "soft infrastructure" such as the integration, standardization, and synchronization of customs and trade regulations, not to mention the entire realm of efforts to secure the actual space of these logistics corridors' (65).

This section of the chapter digs into some of the documents that are trying to organize and develop the knowledge and concepts related to this 'soft infrastructure.' While 'the corridor agenda is increasingly widely adopted by governments, the private sector, and development agencies,' the *Trade and Economic Corridor Management Toolkit*, an extensive study on corridors published by The World Bank in 2014, argues that 'there has been a lack of guidance on how to design, determine the components to include, and analyze the likely impact of corridor projects' (Kunaka and Carruthers 2014, xiii and 1). This is reflected in the heterogeneous array of documents, guidelines, policy papers, and master plans that are produced by nation-states, regional cooperation organizations, and other international bodies worldwide. Only a few studies have attempted to outline a comprehensive understanding of corridors, and these have primarily been related to transport scholarship. However, corridors—whether devoted to physical distribution or data transfer—entail a direct role for the authorities ruling in the spaces they cross or connect, including city councils, regions, states, and international organizations concerned with trade, development, and cross-border activities. At the same time, they produce specific governing dynamics and management agencies whose role is growing as a consequence of the relevance assumed by transport connectivity and logistical infrastructure for global exchanges.

While corridors alter existing patterns of territoriality, they also modify the functioning of existing institutions through the operation of technical standards, governance tools, and financial flows: policy papers, master plans, and international studies are part and parcel of the new 'managementese' that links policy-makers across different fields and sectors (Easterling 2014). The use of logistical concepts such as interoperability, multimodality, or *bottlenecks* is becoming increasingly relevant in the process of channelling political decisions and investment flows in certain directions. The binding of logistical concepts into the larger scheme of corridors and their circulation beyond the realm of transport specialists makes visible the relation of the corridor discourse to power and statecraft. As we will see, corridors, from a subterranean trend in the history of the relation between economy and space, are now emerging as a commonsense reference in discourses of governance and policy-making.

Before attempting a conceptualization of corridors as political forms, however, I will first analyze how corridors are taking the stage in current debates across logistics, particularly as a conceptual tool that is being mobilized with increasing persistence in the policy discussions involving institutional actors as various as China, India, and the European Union, as well as international financial institutions such as the World Bank, the ADB, or the newly formed Asian Infrastructure Investment Bank (AIIB).

As observed by Hans-Peter Brunner (2013), 'economic corridors connect economic agents along a defined geography.' Even if no standard picture of corridors exists, 'we can […] speak of an emerging and fluid concept of what economic corridors are' in the literature and concrete case studies. Brunner highlights how corridors 'are not mere transport connections along which people and goods move' but are in fact 'integral to the economic fabric and the economic actors surrounding it.' As such, corridors must not be understood in isolation but rather 'have to be analyzed as part of integrated economic networks, such as regional and value chains and production networks' (1).

Business studies concede that freight corridors and logistics hubs shape the location decisions of many manufacturers or warehouse and distribution companies, and logistics is becoming the second criterion, after labour's availability and costs, in locating manufacturing and distribution. It is replacing real estate. While, via dedicated software and techniques such as centroid analysis, big companies decide where to invest or locate their activity, new infrastructure and large projects such as highways, railways, and freight corridors add inputs and variables to the software algorithm to create a 'stream of products, services and information moving within and through communities in geographical patterns' (Luttrell 2015a, b; Ghani et al. 2016; Kunaka and Carruthers 2014, 15). Even if a corridor is 'at its core' about 'facilitating supply chains' and connecting locations 'using different modes of transport to link production and distribution centers,' Charles Kunaka and Robin Carruthers (2014), authors for the World Bank of the abovementioned *Toolkit*, concede that corridors also imply connections—between different kinds of actors, spatial settings, operations and decision-making processes—that seek 'to organize production, distribution, and supply to capture regional specialization' (23).

As observed by anthropologist Anna Tsing, the capitalism of the supply chain entails a model for thinking '*both* global standardization *and* growing gaps,' the use of 'preexisting diversity,' and the creation of 'niches and links' among differences. This leads to a global condition in which corporate governance coexists with 'contingency, experimentation, negotiation, and unstable commitments' (Tsing 2009, 151). Against this backdrop, corridors can be understood as the attempt to form logistical institutions, which makes them much more than just physical entities. The implementation of a corridor is generated by deepening infrastructural and economic integration that represents 'the strategic decisions and choices developed and made by firms, municipalities, and governments to attract increased flows of commodities to particular regions' (Kunaka and Carruthers 2014, 23). Such implementation is dependent on not only the 'coalitions' that parties form when attracting investments but also the involvement of institutional actors in a process of legal and territorial reorganization. 'Institutional and economic relationships are part and parcel of a corridor, especially in the presence of competing trade routes' (23). Corridors, in other words, do not simply connect discrete ribbons of infrastructure; they are constitutive of the dynamic political fabric of contemporary capitalism. Therefore, even when they serve a specific and localized need, the political meaning of corridors must be discerned through an exercise of abstraction.

Aspects of these localized, political, and infrastructural characteristics of corridors are reflected in debates between specialists on where corridors originate. If the consulting agency PricewaterhouseCoopers seems to conflate corridors with economic activity, Kunaka and Carruthers mention two alternative bases for corridor projects: the following of historic routes and 'greenfield' developments. While, in the former, corridors correspond to interventions to improve connectivity and infrastructure along existing (historic) routes and connections, 'greenfield' projects create new connections. Nevertheless, as Arnold noted in 2005, existing corridors 'are rarely developed as Greenfield projects' as most of them evolve 'from existing land-based multimodal transport networks' (Arnold 2005, 1). While many greenfield projects have been launched since then, what remains clear is that corridors are conceptually situated between the existing and the new, and their performative role is precisely related to their capacity to translate and shift operations and assemblages into new frameworks, more than simply superimposing new structures over existing ones.

The experience of the New Silk Road, which can be related to projects in Africa as well as South America, demonstrates that the following of historic routes can take unforeseen twists and turns away from original locations. In this sense, the language of the New Silk Road serves a greater role as persuasive discourse than as a connection to real settings. Even when corridor projects try to take advantage of existing geographies and historic legacies, they simultaneously generate new geographies or divert existing paths towards other routes. This is reflected in the shifting relation between so-called path dependency and planning in evaluating the 'transformational impact' of corridors. Otherwise said, a corridor can be, at the same time, 'both a product and an instrument of spatial planning in a country and a region' (Kunaka and Carruthers 2014, 21). The same applies to the economy, where corridors produce a kind of circular dynamic: they serve particular supply chains, but they are also meant to reorient economic and administrative activities in order to integrate them into larger global value chains. In following specific trajectories, corridors are also instruments that can shape the direction and robustness of certain supply chains over others.

The harmonization between these different dimensions and meanings is far from given and is reflected in different definitions that exhibit a 'tension between transport functions, economic functions and spatial functions' (Priemus and Zonneveld 2003, 174). Priemus and Zonneveld, for example, identify three possible ways in which corridors function: as an *infrastructure axis*, as an *economic development axis*, and as an *urbanization axis* (2003, 174). As has already been discussed, the concept of corridor refers both to a 'dynamic space' and a 'productive space,' within which, as Chapman et al. (2003) note, 'infrastructure may need to perform in a variety of different ways' (189). For this reason, they argue, 'conflicts may arise between the range of potential functions of these corridors' at local, national, regional, and global levels (189). That is to say, there is often an unresolved tension between the spatial and institutional dimension of corridors due to their transnational nature in a world marked by national borders, and given that a corridor, by definition, 'also implies physical and linear geographical form more than institutional structure, and homogeneity rather than distinctiveness' (190).

HETERARCHIES AND THE POLITICAL PRODUCTIVITY OF BOTTLENECKS

As evident from the above discussion, corridors are twofold. On the one hand, they are a spatial description of something happening on the ground. And on the other, they are a projection of something to be realized: instruments that anticipate a future through governance tools that synchronize the administrative time, procedures, and standards of bordered institutions. In this way, 'a corridor is [...] a spatial structure for overcoming the fragmentation of legal, institutional, physical, and practical boundaries' (Kunaka and Carruthers 2014, 23–25). In practice, however, Arnold identifies three different kinds of institutional formats that have been used thus far to realize and manage corridors (2005, 36–38):

1. The *disjointed incrementalism model* which promotes improvement along a route classified as a corridor based on local requirements, without a formal corridor organization. As in the case of the North-South Corridor in West Bengal, India, the corridor works as 'a concept around which various projects are developed' and 'exists by virtue of its growing commercial activity rather than through any organizational structure.'
2. The *legislative development model*, such as the TEN strategies in Europe or the Maputo Corridor in Africa, uses regional legislation to design specific routes as corridors and to provide for the harmonization of standards, simplification of cross-border movements, and funding for infrastructure.

3. The *consensus-building institutional model* which uses regional institutions to mobilize stakeholders and push for reforms in the fields of trade facilitation, border-crossing procedures, and infrastructure financing. This is the model closer to the experience of Eastern Europe and Asia.

Corridors seem, then, to develop as a product of economic and administrative activity and without any real and autonomous power. Nevertheless, a closer look to the ways in which the governance of corridors is organized and conceptualized reveals specific forms repeating themselves across different spaces and institutional settings. Formal corridor initiatives often include corridor management bodies whose main activities include planning, prioritizing, financing, advocating for legislative and regulatory reforms, monitoring corridor performance, promoting corridor use, and piloting trade facilitation and logistics reform (Kunaka and Carruthers 2014, 98). This list of functions shows how the managing agencies of corridor-related projects have grown to be much more than technical bodies. They use technicalities to advocate for specific reforms and priorities inside domains that were part of the sovereign capacities of states.

Since most of the transport and development corridors entail a cross-border dimension, these bodies become the connecting entities between political forms and administrative assemblages that have been historically and formally separate, albeit performing similar functions. Operating as a sort of intermediate and organized body of stakeholders in the global economy, these different groups—consisting of technicians, representatives of national governments and international institutions, as well as leading companies in the field of infrastructure, transport, and production—represent the institutional formation of the political economy of logistics. As the different texts and projects analyzed so far in this chapter clearly show, all corridor initiatives make direct reference to the principles and rules set in the existing international free trade agreements developed by the WTO and other institutions. As we will see later, even the new Chinese initiatives make a direct reference to what is today a shared consensus on the priority of connectivity and public-private partnership for economic growth. This is to say that corridor initiatives are part and parcel of an attempt to foster the global neoliberal agenda via the consolidation of a transnational discourse *and* material constitution that finds its basis of legitimization in logistical rationality. Their aim seems to be that of formalizing a third political space between the operations of private capital and political institutions, where these blur in favour of a logistical polity that follows its own temporality, somehow free from the procedures of traditional politics. In this third space, different political systems—whether based on representative politics, state parties, or authoritarian regimes—can meet and share a common ground.

In this common ground 'cooperation,' the ability to build 'coalitions' and 'trust,' are considered even 'more important than geography' (Kunaka and Carruthers 2014, 24–25). Geopolitical considerations coexist with the need to produce a financial *hype* that is necessary to mobilize money. The implementation of an infrastructural corridor is somehow a bet on the capacity to produce a final corridor capable of sustaining investments beyond any specific actor. This also shows the relevance of how different agents can have a role in shaping the formation of a corridor. While policy papers refer mainly to stakeholder forums or institutional settings that bring people together in the building of consensus, other subjects can intervene and influence how a corridor develops. For example, private enterprises or state-controlled companies can intervene in operations such as the development of physical or financial infrastructure that can be more effective than forums or official initiatives in imposing a corridor agenda. In terms of the definition of corridors, we can thus observe a dynamic of 'competitive alignment' between different subjects, both public and private. This is reflected in the concept of 'heterarchy' which is defined by de Vries and Priemus as a 'self-organized steering of multiple agencies, institutions, and systems that are operationally autonomous from one another yet structurally coupled as a result of their mutual interdependence' (2003, 226). Given the loose political coordination at the transnational level, the realization of *core points* along projected routes can be used as a bridgehead to create the economic and political tension that supports corridor projects. In this way, the politics of corridors meshes different economic spatial schemes—a view of corridors as the natural development of existing trade and exchange, or as an instrument for planning something that does not currently exist—by creating the environment and levers for the restructuring of space according to logistical projections. Logistical connectivity produces splintering effects on the existing organization of space while creating, at the same time, its own paths of territorialization along infrastructure, trade, and supply routes.

While automation and the use of GPS data make possible the measuring of corridor performance, the assessment of the footprint of logistics in the overall economy remains more complex. The issue of monitoring how logistics performs, even for industry itself, remains tricky, although different survey models have been used to target single-company performance or corridor-specific functions. Yet the industry seems to agree that the most important characteristic associated with '"logistics friendly" countries' is the level of sophistication of their service, which allows manufacturers to outsource logistics to third-party providers, increase their competitiveness 'and focus on their core business while managing more complex supply chains' (Arvis et al. 2014, 3). The role of logistics is widely considered to be crucial in order to connect global supply chains and thus take advantage of other parameters such as low labour costs or abundant natural resources. If physical infrastructures are obviously important, there is a growing understanding that infrastructure alone is insufficient in the development of better performance of the whole 'transit system.' This is indeed incorporated in a 'transit regime' which includes all the 'infrastructure, legal framework, institutions, and procedures serving trade corridors' necessary to make possible 'the movements of goods from their origin (often a seaport) to their destination (such as a clearance center in the destination country)' (Arvis et al. 2014, 33). The formation of transnational 'transit regimes' requires high degrees of cooperation at the level of governance, regulation, and standardization, and such coordination reveals the political layering of the technical practices central to logistics.

Commenting on the EU's TEN policy, Opitz and Tellman observe that the 'conceptual enfolding of the physicality of infrastructure' becomes especially apparent when considering the concept of *bottleneck*. While the concept may seem simple, *bottleneck* 'turns out to be a hybrid term that problematizes all sorts of different impasses in interconnectivity' by turning 'rather different issues into a similar problem: that of a barrier, which is to be erased by a physical-cum-economic connectivity' (Opitz and Tellman 2015, 181–82). Together with other terms in the logistical discourse, the use of the term *bottleneck* is increasingly associated with a space of political experimentation. The definition of any point on a map as a *bottleneck* depends on causes that are both evaluated from past and present experience and foreseen by future projections regarding the performance of the related network. Different types of evaluations can enter the process: market conditions, traffic and demand data, environmental conditions, technical standards, and geo- and socio-political constraints. The consulting agency PricewaterhouseCoopers explains that while the term *bottleneck* is in common use at the policy level, 'it becomes more problematic if a precise definition is required of what exactly is and is not a bottleneck and, more particularly, where bottlenecks might potentially occur in future' (PricewaterhouseCoopers 2011, 70). In practice, 'bottleneck identification will need to be pragmatic,' (9) and its definition can be considered as a performative act that lies between the technical and the political and across past conditions and future settings. This ambiguity makes the exercise of identification and definition of bottlenecks resonate with the cartographic practice of defining the world on a map (Mezzadra and Neilson 2013, 27–59). As PricewaterhouseCoopers explains, 'a bottleneck is [...] degradation in quality of service relative to some norm' but 'what the norm is,' as well as 'what constitutes a degradation of service of sufficient severity to justify "bottleneck" status,' can often be 'a matter of judgment.' As the agency concludes, 'there is no principled basis for drawing the boundary between bottleneck and no bottleneck. [...] The best that is likely to emerge is some kind of expert consensus about what profile of characteristics might reasonably permit a location on a grid to be regarded as a bottleneck' (PricewaterhouseCoopers 2011, 70).

Together with *bottleneck,* other concepts driven by logistics, such as 'intermodality' and 'interoperability,' reveal the impact on governance processes of logistical operations. While these concepts refer primarily to the technical problem of making possible and smooth the connections between different transport systems and networks, their role inside corridors is to 'reduce fragmentation of jurisdictional, infrastructural, procedural, management, and other boundaries [through] harmonization of laws, institutional frameworks, norms, standards, and practices based on internationally agreed standards' (Kunaka and Carruthers 2014, 19).

ASIA'S ERA OF INFRASTRUCTURE AND THE POLITICS OF CORRIDORS

There is a long history of efforts to connect Asia along transport corridors. Plans to develop an Asian Highway (AH) and the idea of a Trans-Asian Railway (TAR), connecting Singapore with Istanbul, began to circulate in the 1950s and 1960s. Yet it was only in the 1990s, in the wake of liberalization processes and the development of transnational value chains, that the demand for physical connectivity to support export-led growth strategies and fragmented production networks grew extensively. In 1992, the United Nations Economic and Social Commission for Asia and the Pacific (UNESCAP) launched the ALTID project (Asian Land Transport

Infrastructure Development), promoting the realization of the AH and TAR. Since then, the AH and the TAR have become 'major building blocks of the development of an international integrated intermodal transport system in Asia and beyond' (De 2012, 157).

The argument behind the promotion of pan-Asian infrastructures resonates with mainstream discourse on development that includes infrastructure and logistical connectivity as the foundation for economic growth in a global world. This discourse considers the development of regional demand and consumption as key to overcoming the crisis that originated in 2008, and regional connectivity—including the development of 'soft infrastructure'—as essential for its achievement (De 2012; Hoontrakul et al. 2014, 167–88). Following intense negotiation, intergovernmental agreements on the AH and TAR networks were adopted in 2004–2005, involving countries from Iran to the ASEAN nations, including India and the PRC (159, 174). It is against this background that China, after years of intense investment in logistical infrastructures across the world, officially launched the New Silk Road or Belt and Road Initiative (National Development and Reform Commission 2015).

An article published by the *Financial Times* in 2015 observed that the hype created by the Belt and Road discourse has not yet been met with concrete projects (Hornby 2015). Only a few investments, including in Pakistan and South Africa, have been directly connected to this plan. The article specifies that while the Asian Infrastructure Investment Bank (AIIB) 'could fill the gaps once it is up running,' it is impossible to make serious provisions due to lack of details. A further observation is that China, like Western Europe after the fall of the Soviet Union, is now engaged in the development of 'investment protocols' to bridge legal and corporate norms between itself, Central Asian countries, and other countries (Hornby 2015). Yet, one should pay attention to the words used by Silk Road enthusiast Zhao Changhui, chief risk analyst at China Export-Import Bank, who stresses that the Belt and Road Initiative is more relevant as 'a new method of development for China and the world' than for the numbers it involves (Hornby 2015). The New Silk Road is, in fact, a particular articulation of the politics of corridors that is simultaneously distanced from, and intertwined with, the forms analyzed by Arvis, Kunaka, and Carruthers.

The plan presented in the Chinese government's *Vision and Actions on Jointly Building a Silk Road Economic Belt and the 21st Century Maritime Silk Road* mixes the historical 'Silk Road Spirit' with what can be called a 'supply chain vision' that is built on the development corridor scheme (National Development and Reform Commission 2015). The document calls for 'policy coordination' and 'new models of international cooperation and global governance.' It has two aims: first, to guarantee 'unimpeded trade' without political or infrastructural obstacles; second, to bring development to inner China and the landmass of Central Asia through a coordinated and planned vision, where the new "green" components of China's industry peek out from between the lines. The key aims read as follows:

1. Improve the division of labor and distribution of industrial chains by encouraging the entire industrial chain and related industries to develop in concert;
2. Establish R&D, production and marketing systems;
3. Improve industrial supporting capacity and the overall competitiveness of regional industries; increase the openness of our service industry to each other;
4. Explore a new mode of investment cooperation, working together to build all forms of industrial parks such as overseas economic and trade cooperation zones and cross-border economic cooperation zones, and promote industrial cluster development;
5. Promote ecological progress in conducting investment and trade, increase cooperation in conserving eco-environment, protecting biodiversity, and tackling climate change. (National Development and Reform Commission 2015)

The text also implies the 'core' concept found in other corridor projects, referring to a string of 'core cities' and 'key economic parks' along the route. While the OBOR initiative has its own map, its reach can potentially extend well beyond the selected areas and reach any point on earth, where a *bottleneck* must be removed or new infrastructure is needed. After a first round of investments along the way, including, for example, the COSCO's intervention in the port of Piraeus, Greece (Grappi 2015), President Xi Jinping and Prime Minister Li Kequiang announced the creation of the AIIB in October 2013. The AIIB begun operations in 2015 and is seen as a potential competitor with present international financial institutions such as the International Monetary Fund, the World Bank, and the ADB. Maybe for this reason, the *Toolkit* published by the World Bank makes no mention of the role of China in infrastructure development, even if it was discussed at least since 2011 (Brown 2011). Casting an eye over the almost 60 members of the AIIB and the projects approved so far reveals a different story. At the time of writing, seven projects were already listed on the AIIB (2016) website involving Kazakhstan, Pakistan, India, Bangladesh, Tajikistan, and Indonesia. Of the seven, three are cofinanced with the World Bank, one includes the ADB, and one involves the European Bank for Reconstruction and Development—for an intervention in Corridor 3 of the CAREC initiative which in turn is an ADB initiative. These arrangements confirm the character of 'competitive alignment' on infrastructural projects mentioned before.

Deborah Cowen, in an analysis of the leading role of China in corridor and gateways projects, observes that 'rather than a world factory, China might be better conceptualized as a logistics empire' (Cowen 2014, 67–68). This 'empire' is nevertheless peculiar. While it surely reflects the economic priorities of China, including the relevance of the logistics industry in the country, it also reflects a friction between national interests, the ideologically cooperative form of infrastructure, and the strategic gaze of logistical power (Grappi 2016, 153–74; Neilson 2012). Gordon Orr, senior adviser at the McKinsey Consulting Agency, suggests that the Chinese economy 'is today made up of multiple sub economies' of which 'some are booming, some are declining, some are globally competitive, and others are fit for the scrap heap' (Orr 2016). Splintered infrastructures and urban congestion, together with the dependency on ports in the East due to lack of connections in the West, are among the biggest threats to Chinese performance in the future, and one of the main problems China already has to face. This explains the stress placed by the government on infrastructure. Yet logistical corridors are more than simply technical answers to economic questions. In fact, as we have seen before, through the corridor logic they are producing new political dynamics that link the economy and institutions as well as market behaviour and government planning. Even the famous Jing-Jin-Ji Project (Jing for Beijing, Jin for Tianjin, and Ji for the ancient name of Hebei Province), which is set to realize the biggest urban corridor in the world, should thus be understood as a way to address the disorderly urban growth of the region by government planning, instead of as a 'fictional' projection of Chinese gigantism, as is often portrayed in the West (Johnson 2015). In the PRC's 13th Five Year Plan, it is not by chance that behind the language of decentralization, more centralization is to be expected from the implementation of the Belt and Road. This centralization may nevertheless take unexpected twists since it implies the implementation of the corridor logic on a wider scale, involving both the entire Chinese territory and Asia.

What China is promoting is thus something more than a new role; it is a vision, a sort of 'logistical Confucianism' that has absorbed the principles of the 'logistical revolution' and now invites all regional and global powers to cooperate in the name of economic growth (Grappi 2016, 173; Allen 1997). The logistical nature of this vision is reflected in what is considered to be the precondition: infrastructure. And, as Professor Shi Ze from the government think tank China Institute of International Studies articulates:

> In accordance with the existing economic foundation and condition, and the pre-conditions of our cooperation, our efforts must first be put to energy resources, the transportation grid, electricity systems, communications networks, other such basic infrastructure platforms, and the networking together of such platforms. There is a saying in China, 'to develop wealth, you have to first build roads.' The development of the corridor's economy can only prosper when human resources, logistics and economic flow have all been brought on-line and integrated. These basic conditions must be there. (Shi 2014)

Asked about the presence of many 'geopolitical bombs' along the Belt and Road in the course of a panel on 'Asia's Era of Infrastructure' held in Davos, the president of AIIB Jin Liqun stated that the bank will not fund projects involving disputed areas—which is questionable—but that all countries should understand that connectivity is important. He added that a criticism of the role of China inside the AIIB means a criticism over any of the multilateral banks already existing, giving that in all cases there is a country that contributes with more funds and more initiative (World Forum 2016).

Economy, Space, and History: Corridors as Political Forms in the Logistical Century

The ways in which the logistical centre of gravity of Chinese strategies is affecting the political formation of new governance instruments and institutional apparatuses shows commonalities, with other processes and experiences. As Deborah Cowen notes, 'it was geographers who were debating (and mapping) corridors and gateways four decades ago in a disciplinary conversation [on urban systems] that has largely disappeared since' (Cowen 2014, 63). The reference to this kind of debate is the starting point of a special issue of the journal *Transport and Geography* dedicated to the governance of corridors (Priemus and Zonneveld 2003). What is relevant for our discussion is how the upgrading of corridors from essentially an urbanity problem to a larger concept related to economic planning and development has led, in Europe, to the definition of megacorridors or, later, trans-European corridors. These concepts stressed the role of the corridor as a 'planning concept' used by town planners since the nineteenth century, for example, with the Ciutat Lineal concept of Spanish planner Soria y Mata. In their introduction to the special issue, Priemus and Zonneveld argue that megacorridors 'should not be conceived of as an entity occupying physical space' because they:

> comprise the arena within which an attempt must be made to arrive at an integration of a multitude of social interests. In this sense, a corridor is neither a sectorial nor a spatial concept, but rather the indication of a challenge: that of improving the governance of infrastructure and area development. (Priemus and Zonneveld 2003, 176)

Even the European Commission understands corridors as a comprehensive planning concept for spatial development with cross-governance implications—a conception that resonates in the *Vision and Actions* presented by the Chinese government to qualify the Belt and Road Initiative (CEC 1999, 36; National Development and Reform Commission 2015).

As already mentioned, when looking at corridors as a planning concept, two different implications must be considered. First, different models for corridor planning and corridor management exist, having different implications regarding the involvement of the existing institutional settings. Second, the design of corridors is the result of different streams of mapping and envisioning the economic spatial structure. Corridors can be the result of economic processes of concentration along certain routes and in certain kinds of clusters. On the other hand, the planning of corridors can create new economic flows. It is conceptually relevant to note that, in the discussion around corridors, a thinking space is created whereby distinctions—between the economic and the institutional, the public and the private, the national and the supranational, and the local and the global— tend to blur and even vanish. This highlights the inner nature of corridors as forms that stretch economic and political processes across and beyond the established maps, allowing peculiar forms of politics and economy to temporarily crystallize.

From an economic point of view, the spatial distribution of economic activities after the 'logistical revolution' took the form of firm-centred corridors along the territorial advancement of supply chains. The flows created by the extension of the supply chain beyond the original reach of each enterprise's activity created a grid of connections within the form of loose corridors. Containerization helped to centralize these connections at the transnational level because carriers connected flows originated by different firms. In this way, containerization has helped forge powerful networks and pushed for a geographical restructuring of ports and related infrastructures. The growth of so-called third-party logistics services and their concentration around a few key global players is another factor that pushes the organization of otherwise disorganized inland transport activities, matching them with the internal organization of the assembly (or extraction) line.

Considering the logistical politics around corridors gives us the analytical capacity to connect, organize, and valorize a large spectrum of unbalances: between regions, within regions and countries, between labour conditions and productivity, around stability and mobility, and between extreme performances and sudden crisis. These dynamics are of great interest for a theory of the state form and its variations. A useful reference here is that of the 'global state': a concept proposed by Italian scholar Maurizio Ricciardi to describe the contemporary state as a 'critical actor' which derives political legitimacy from the capacity to act across different fields, rather than from the modern form of the social contract (Ricciardi 2013, 13). This state can assume different shapes but is always marked by a fundamental 'incompleteness' regarding its sovereign capacities and legitimacy (19).

This essay attempts to connect the scholarly debates and technical knowledge on corridors, with the dimension of logistical operations, to show how corridors may emerge as political formations against this background. If we consider the trajectories and transformations of the state and sovereignty, Lauren Benton (2010) describes the colonial expansion of European states as taking the form of 'corridors and enclaves' and argues that these should be considered part and parcel of the wider history of modern sovereignty, and as positioning the territorial state as a specific and situated experience. Even if the state form has in the twentieth century subsumed all other forms of political and institutional formations, leading to its proliferation, Saskia Sassen describes it as a particular assemblage of already-existing capacities into the institutional fabric of the territorial state (Anghie 2004; Sassen 2006). Does this mean that the corridors we are referring to are nothing new, or that they are just the contemporary expression of a longer history? The answer is twofold.

On the one hand, as we have seen, the history of sovereignty and of the state is marked by the corridor form. On the other, however, the global politics of corridors is a global institutional form that is organized and conceptualized around the discourse of logistics and its operability. As the logistical politics of corridors is generating a new dynamic in governance that is creating its own instruments and political discourses, these are penetrating and imposing themselves in different ways inside the many sites of global politics. Unlike the 'snaking pattern' of jurisdiction traced by Benton, logistical mapping is not about control and authority building, but about projecting and pushing further—and rendering and visualizing hidden integration lines along global value chains. Even when geopolitical enlargement is the focus, this happens within a global grid of economic, financial, and commodity flows that reflect the conceptual shift produced by the 'logistical revolution.' These processes refer to different relations between power, politics, and space, rather than a global political authority over faraway territories, and the entity that is supposedly enlarging is, rather, an always uncompleted and transforming actor (Benton 2010; Cowen 2014, 64; Ricciardi 2013; Mezzadra and Neilson 2014).

If Benton distinguishes two kinds of corridors—as places where the movement of goods and people occur and as instruments in the process of imagining and constructing sovereignty—global logistical power sees corridors in still another way. In fact, they are conceived both as concrete networks of infrastructure and as governance structures that constitute centres of power. There is thus a difference between an understanding of corridors as the expression of fragmented sovereignties and corridors as political forms. Rather than consider logistics and corridors as functional instruments of a particular sovereign power, which is affected by their form but remains nevertheless the focus of the analysis, as in the conceptualization of the 'logistical State' advanced by Henri Lefebvre, I shift the gaze to corridors in order to observe the political dimensions of logistics (Lefebvre 2009; Toscano 2014; Grappi 2016, 25–28, 103–30).

Conclusion

While corridors entered the economic and political debate as part of the promotion of the market economy that defines connectivity, accessibility, and infrastructure as a *conditio sine qua non* for economic growth and competitiveness in a world economy, the creation of development corridors is also related to the idea that balanced and sustainable economic growth can bring back the role of the state and its institutions. The very construction of infrastructure is a way to boost economies and create jobs. Large corridor plans can thus be considered a mix of neoliberal austerity policy and Keynesian demand stimulation practices (Albrechts and

Coppens 2003, 217–18). More widely, the 'politics of corridors' advanced in this essay has to do with two fundamental dimensions: the leading role of logistical operations and logistical rationality in the drawing, planning, and implementation of its material imagination; and the formation of new tools of governance that reshape the relation between assemblages of territory, authority, and rights. In creating its own logic and governing bodies, the politics of corridors is an emerging form of 'global power' where government and governance intersect (Schiera 2013, 34). By saying this I am not suggesting that the aim is to govern the global as a unique superstructure, but rather that the politics of corridors is a distinctive form of government *vis-à-vis* the global. As many have already observed, a distinctive feature of the global is that this does not imply the disappearance of previous forms. Nevertheless, the state is transformed by the politics of corridors. This politics should not be evaluated merely by its success, or by judging the adherence between master plans and their practical realization, but from its capacity to orient debates and decisions, and from the simultaneous dynamics of cooperation and competition it generates between existing institutions. We can in this way appreciate how the relation between regional and global dynamics reshapes not only spatial relations but also conceptual nomenclatures. If the world meets in Baku, as we say opening this essay, this means that we need to look beyond the matrix of pure economic interests that merge in the area: from resource extraction to tax breaks, from geopolitical realignments to the possibility to open new markets and find fresh labour force. What we need is to analyze how these dimensions intertwine, transform, and reshape the ways in which political power is being achieved and practised today, and how it pursues its unstable legitimacy. The focus on corridors suggests that the spatial reorganization they entail is accompanied by a transformation of the role institutions do and will play inside neoliberalism.

The politics of corridors, constructed by adopting the political discourse of logistics, is taking a step forward in terms of the political reorganization achieved through zones as well as in the servicing of global supply chains. It is creating enclaves inside the territories of the state and fragmenting its space, while at the same time reinforcing the role of the state and redefining its sovereign capacities. The need to understand more deeply how the different politics of corridors overlap, conflict, and diverge in specific places on the ground, and in relation to different institutional settings—as well as the trajectories of resistance and frictions produced through the emergence of the corridor form and its variations—suggests a further direction for research. By proposing the concept of the 'politics of corridors,' my aim is to move our gaze from the single infrastructure, the singular point where logistical operations hit the ground, to the larger political rationality and variegated geographies of logistical power that are behind the realization of these projects, and behind their subsequent reorganization of political and social spaces. How to confront the politics of corridors, and how to organize the subjectivities that it encounters, generates, and puts into motion, is a major task for contemporary political theory.

REFERENCES

AIIB, Asian Infrastructure Investment Bank. Accessed 30 August 2016. http://euweb.aiib.org/

Albrechts, Louis and Coppens, Tom. 2003. 'Megacorridors: Striking a Balance Between the Space of Flows and the Space of Places.' *Journal of Transport Geography* 11, no. 3: 215–24.

Allen, W. Bruce. 1997 'The Logistics Revolution and Transportation.' *Annals of the American Academy of Political and Social Science* 553: 106–16.

Anghie, Antony. 2004. *Imperialism, Sovereignty, and the Making of International Law*. Cambridge and New York: Cambridge University Press.

Arnold, John. 2005. *Best Practices in Corridors Management*. Washington DC: The World Bank.

Arvis, Jean-François, Daniel Saslavsky, Lauri Ojala, Ben Shepherd, Christina Busch, and Anasuya Raj. 2014. *Connecting to Compete 2014. Trade Logistics in the Global Economy. The Logistics Performance Index and Its Indicators*. Washington DC: The World Bank.

Benton, Lauren. 2010 *A Search for Sovereignty: Law and Geography in European Empires, 1400–1900*. Cambridge University Press: Cambridge.

Brown, Kevin. 2011. 'Full Speed Ahead on New Silk Road.' *Financial Times*, 9 November 2011.

Brunner, Hans-Peter. 2013. 'What is Economic Corridor Development and What Can It Achieve in Asia's Subregions?' *ADB Working Paper Series on Regional Economic Integration* 117.

CEC. 1999. ESPD: European spatial development perspective. Towards balanced and sustainable development of the territory of the European Union. Luxembourg: Office for Official Publications of the European Communities. Accessed 30 August 2016. http://ec.europa.eu/regional_policy/sources/docoffic/official/reports/pdf/sum_en.pdf

Chapman, David, Dick Pratt, Peter Larkham, and Ian Dickins. 2003. 'Concepts and Definitions of Corridors: Evidence from England's Midlands.' *Journal of Transport Geography* 11, no. 3: 179–91.

Cowen, Deborah. 2014. *The Deadly Life of Logistics: Mapping Violence in Global Trade*. Minneapolis: University of Minnesota Press.
De, Prabir. 2012. 'Does Governance Matter for Infrastructure Development? Empirical Evidence from Asia.' *Journal of Infrastructure Development* 4, no. 2: 153–80.
de Vries, Jochem, and Hugo Priemus. 2003. 'Megacorridors in North-West Europe: Issues for Transnational Spatial Governance.' *Journal of Transport and Geography* 11, no. 3: 225–33.
Easterling, Keller. 2014. *Extrastatecraft: The Power of Infrastructure Space*. London and New York: Verso.
Hoontrakul, Pongsak, Christopher Balding, and Reena Marwah, eds. 2014. *The Global Rise of Asian Transformation: Trends and Developments in Economic Growth Dynamics*. New York: Palgrave Macmillan.
Ghani, Ejaz, Arti Grovewe Goswami, and William R. Kerr. 2016. 'Highway to Success: The Impact of the Golden Quadrilateral Project for the Location and Performance of Indian Manufacturing.' *The Economic Journal* 126, no. 591: 317–57.
Grappi, Giorgio. 2015. 'Logistics, Infrastructures and Governance after Piraeus: Notes on Logistical Worlds.' *Logisticalworlds.org*, 4 August 2015. http://logisticalworlds.org/blogs/logistics-infrastructures-and-governance-after-piraeus-notes-on-logistical-worlds
Grappi, Giorgio. 2016. *Logistica*. Roma: Ediesse.
Hornby, Lucy. 2015. 'China's "One Belt One Road" Plan Greeted with Caution.' *Financial Times*, 20 November 2015. http://www.ft.com/intl/cms/s/2/5c022b50-78b7-11e5-933d-efcdc3c11c89.html#axzz3x8bAQFOa
Kunaka, Charles, and Robin Carruthers. 2014. *Trade and Transport Corridor Management Toolkit*. Washington, DC: World Bank.
Lefebvre, Henri. 2009. *State, Space, World: Selected Essays*, edited by Neil Brenner and Stuart Elden. Translated by Neil Moore, Neil Brenner, and Stuart Elden. Minneapolis: University of Minnesota Press.
Luttrell, Bill. 2015a. 'Freight Corridors and Logistics Hubs Shape the Location Decision.' *Area Development*. Accessed 30 August, 2016. http://www.areadevelopment.com/logisticsInfrastructure/Intermodal-Sites-Q1-2015/site-selection-process-supplychain-optimization-linked-74421.shtml
Luttrell, Bill. 2015b. 'Logistics Technology: Locations That Improve Supply Chain Optimization.' *Site Selection Magazine*. July 2015. Accessed 30 August 2016. http://siteselection.com/issues/2015/jul/logistics-technology.cfm
Mezzadra, Sandro and Brett Neilson. 2013. *Borders as Method, or, the Multiplication of Labor*. Durham and London: Duke University Press.
Mezzadra, Sandro and Brett Neilson. 2014. 'The State of Capitalist Globalization.' *Viewpoint Magazine* 4. https://www.viewpointmag.com/

Mooney, Turloch. 2015. 'Baku Port Builds for Opportunities in China-Europe Overland Links.' *IHS Fairplay* 26 November 2015. Accessed 30 August 2016. http://fairplay.ihs.com/ports/article/4257456/caspian-sea-port-of-baku-builds-for-opportunities-in-china-europe-overland-links
National Development and Reform Commission. 2015. *Vision and Actions on Jointly Building Silk Road Economic Belt and 21st-Century Maritime Silk Road*. Ministry of Foreign Affairs and Ministry of the People's Republic of China.
Neilson, Brett. 2012. 'Five theses on Understanding Logistics as Power.' *Distinktion: Scandinavian Journal of Social Theory* 13, no. 2: 322–39.
Orr, Gordon. 2016. *What Might Happen in China in 2016*. McKinsey & Company. Accessed 26 April 2017. http://www.mckinsey.com/insights/strategy/what_might_happen_in_china_in_2016
Johnson, Ian. 2015. 'Chinese Officials to Restructure Beijing to Ease Strains on City Center.' *The New York Times*, 11 July 2015. Accessed 30 August 2016. http://www.nytimes.com/2015/07/12/world/asia/china-beijing-city-planning-population.html
PricewaterhouseCoopers. 2011. *North-South Interconnections: Market Analysis and Priorities for Future Development of the Electricity Market and Infrastructure in Central-Eastern Europe under the North-South Energy Interconnections Initiative*. Final Report for the European Commission. https://ec.europa.eu/energy/sites/ener/files/documents/2011_wg_north_south_interconnections.pdf
Priemus, Hugo and Wil Zonneveld. 2003. 'What are Corridors and What are the Issues? Introduction to Special Issue: the Governance of Corridors.' *Journal of Transport Geography* 11, no. 3: 167–77.
Sassen, Saskia. 2006. *Territory, Authority, Rights: From Medieval to Global Assemblages*. Princeton: Princeton University Press.
Ricciardi, Maurizio. 2013. 'Dallo Stato Moderno allo Stato globale. Storia e trasformazioni di un concetto.' *Scienza & Politica* 45, no. 48: 1–19.
Opitz, Sven and Ute Tellman. 2015. Europe as Infrastructure: Networking the Operative Community. *South Atlantic Quarterly* 114, no. 1: 171–90.
Schiera, Pierangelo. 2013. *Dal potere legale ai poteri globali. Legittimità e misura in politica*. Bologna: Dipartimento di Scienze Politiche e Sociali. Accessed 14 March 2016. http://amsacta.unibo.it/3655/1/Quaderno_Schiera_1.pdf
Shi, Ze. 2014. '"One Road and One Belt" and New Thinking with Regard the Concepts and Practice.' Lecture at the Shiller Institute, Frankfurt, 18 October 2014. Accessed 30 August 2016. http://newparadigm.schillerinstitute.com/media/one-road-and-one-belt-and-new-thinking-with-regard-to-concepts-and-practice/
Skyscraper City. 2016. Thread Baku | Projects and Construction. Accessed 30 August 2016. http://www.skyscrapercity.com/showthread.php?t=610923

Sulayem, Ahmed bin. 2013. 'Port-Centric Logistics Corridors are Future.' *Khaleej Times* 25 February 2013. Accessed 30 August 2016. http://www.khaleejtimes.com/article/20130225/ARTICLE/302259936/1002
Toscano, Alberto. 2014. 'Lineaments of the Logistical State.' *Viewpoint Magazine* 4. https://www.viewpointmag.com/
Tsing, Anna. 2009. 'Supply Chains and the Human Condition.' *Rethinking Marxism* 21, no. 2: 148–76.
World Forum. 2016. Asia's Era of Infrastructure. Accessed 30 August 2016. https://www.weforum.org/events/world-economic-forum-annual-meeting-2016/sessions/asia-s-era-ofinfrastructure

CHAPTER 10

Logistics of the Accident: E-Waste Management in Hong Kong

Rolien Hoyng

Promotional materials by the Hong Kong SAR government depict Hong Kong as an 'ICT Hub,' making reference to Internet speeds that are the world's fourth fastest and a mobile penetration rate of over 200 per cent (Office of the Government Chief Information Officer 2017). However, environmentalist organizations such as Greenpeace and Friends of the Earth reverse the purport of such city branding campaigns by naming Hong Kong an 'E-waste Hub,' not an ICT Hub. Rather than 'virtual' connectivity, their reports highlight Hong Kong's entanglement in illegal circuits of e-waste (electronic waste) disposal, recycling, and trade (Friends of the Earth 2011; Greenpeace 2003). The *E-Trash Transparency Project* report entitled *Disconnect: Goodwill and Dell, Exporting the Public's E-Waste to Developing Countries* (Puckett et al. 2016) by the advocacy group the Basel Action Network (BAN) goes as far as to claim that Hong Kong has replaced China as a 'pollution haven.' Following BAN's independent research, out of the sample of 200 items that they enhanced with GPS (Global Positioning System) trackers and disposed of through vari-

R. Hoyng (✉)
School of Journalism and Communication, Chinese University of Hong Kong, Shatin, Hong Kong

© The Author(s) 2018
B. Neilson et al. (eds.), *Logistical Asia*,
https://doi.org/10.1007/978-981-10-8333-4_10

ous reputable recyclers in the US, 65 were exported and 37 arrived in Hong Kong. They ended up in what BAN deems in all likelihood to be facilities that do not meet legal requirements. Not only imported e-waste but also e-waste generated from within Hong Kong itself is reported to circulate in unchecked ways. According to estimates from 2011, no more than 17 per cent of used computers were designated for governmental disposal facilities due to the salvaging activities of informal collectors, traders, and repairers (Chung et al. 2011, 544).

This chapter focuses on the logistical practices, software, discursive categorizations, regulatory techniques, and normative narratives that manage the mobility and materiality of e-waste in Hong Kong's formal and informal circuits. In practice, e-waste appears to be an ambiguous denomination and its mobility is not strictly surveilled. As Hong Kong doubles as a knowledge-driven ICT Hub and a hub for e-waste recovery activities, the city problematizes narratives of 'e-waste dumping' that assume singular trajectories of e-waste flowing from developed regions to developing regions, alongside narratives of the linear transition from product to waste (Lepawsky 2015). Hong Kong hosts advanced e-waste recycling industries that do the logistical work of arranging highly complex and geographically distributed processes of recycling. Yet, at the same time, environmentalist organizations argue that Hong Kong has not only functioned as a port for often illegal global e-waste exports into mainland China, but that, in reaction to stricter regulations and changing conditions in the mainland, the city has become a site where imported e-waste accumulates and lingers. Illegal storage and processing takes place in e-waste yards in rural districts such as Yuen Long in the New Territories. Forming nodes in informal and illegal circuits of e-waste recycling, the less-hyped parts of Hong Kong host remarkably vibrant scenes of e-waste repair, refurbishment, and trade. These scenes unfold at street markets such as Sham Shui Po's hawkers' market, bazaars such as Chung King Mansion and Sin Tat Plaza, and warehouses situated in the former industrial district of Kwun Tong.

Hong Kong's situation raises the following questions: How do legal and normative discourses as well as logistical practices and technologies manage e-waste's mobility and transformation? At what points in e-waste's circulation are discursive categorizations upheld, rendered ambiguous, or obscured altogether? What are the stakes of spatial concepts such as 'territory' and 'region' in the analysis of e-waste mobilities and transformations?

This chapter argues that e-waste possesses an elusive, transient quality that is paradoxically both constructed as such through actual governance *and* emergent in practice in an uncontrolled fashion that undermines social power and territorial control. The fact that e-waste's mobility and transformation challenge categorization and oversight does not just signify the failure of governance but also the reconfiguration of power and its techniques.

In terms of methodology, this study elaborates three foci: (1) discourses of e-waste governance and regulation, (2) logistical practices and their technologies, and (3) transient matter. First, in order to analyze discourses, norms, and categorizations as regulatory techniques, I consulted policy documents and technical information pertaining to legal regulations, licences, permits, and certificates. I further conducted semi-structured, in-depth interviews with particular institutional actors including the Environmental Protection Department (EPD) of the Hong Kong SAR government as well as several NGOs dealing with e-waste, namely Caritas, St. James' Settlement, Friends of the Earth, and Greenpeace. Second, in order to analyze logistical practices and their technologies, I consulted white papers and promotional material for software in reverse logistics. With the help of a research assistant, I conducted ethnographic observation and interviews with sellers, distributors, repairers, and waste collectors at the following markets for electronics secondhand trading and repair: Chungking Mansions, Sin Tat Plaza, and Sham Shui Po Market in Hong Kong as well as Huaqiangbei in Shenzhen, mainland China. Last, visiting formal and informal recycling sites, I approached e-waste as transient matter that evolves into something else, rendering its designation as 'waste' temporary and relative rather than absolute. I focused on calculations and visions of e-waste's inherent potential for multiple futures and tried to get a sense of e-waste's recalcitrant material agency by eyeing its capacity to generate unanticipated, disruptive situations, if not its becoming a persistent obstacle, hazard, or nuisance.

The following section theorizes waste in terms of the binary of order/disorder, which underlies much of the critical literature on waste, in connection to questions of power. The subsequent section looks at discourses and techniques of e-waste governance, especially the product/remainder binary and what I call narratives of 'magical circuits.' I proceed by investigating the implications of e-waste's ambiguous status as product *and* remainder for the possibility of effective governance and territorial control. The last section discusses logistical practices and their technologies in terms of the (dis)orders they constitute and the shifting modalities of power and spatial orders they inform.

Order/Disorder and Power

What is waste? In theories oriented on 'human' society, waste often exists primarily as a negative. In capitalist systems of value, as well as critiques of capitalism, waste tends to be simply that which no longer bears any value. Hence waste refers to capitalism's useless overproduction and unacknowledged, accidental by-products that escape market valuation yet may inflict externalized costs and harm. Seeking a diversion from green critiques of capitalism, while remaining loyal to Marxian principles, Jason W. Moore (2015, 1–32) questions the reliance on a nature/society divide that reduces nature to either a malleable resource to be exploited or some sort of dump that absorbs externalities. The problem he finds is that, in such critiques, nature remains inconsequential for the further development of capitalism and, by extension, the social order itself. Reckoning with a yet different set of axioms, new materialist critiques of societally and human-oriented approaches object that the definition of waste as residue fails to address the agency of matter. Following Bennett (2010), such approaches ignore matter's unstable yet powerful effectivities that are not epiphenomenal to society's operation but form the very foundation of human-nonhuman realities. Materialist approaches endow matter with recalcitrance and vitality. Parikka (2015) draws from Deleuze and Guattari's critique of hylomorphist assumptions that immaterial forms shape matter. He mobilizes the geocentric concept of the metallurgical to call for attention to the potentialities inherent in waste as matter. In line with Jackson's (2014) take on 'broken world thinking,' materiality is always on the verge of disintegrating and transiting into something else.

Whereas materialisms of Bennett's line of thinking designate recalcitrant matter a foundational and ontological status, prevailing critiques of capitalism consider it epiphenomenal in relation to primal social forces (Gille 2010, 1054). Though supposedly contrary endeavours, both these approaches adhere to a binary of order/disorder. Waste tends to equate to disorder, undermining human intentionality and social models of utility. This association of waste with 'disorder' also underpins much of the moral condemnation in critical activist discourse pertaining to e-waste dumping.

This chapter sets out to rethink the order/disorder binary in connection with questions of power and governance. One critique of this binary is that it prevents questions being raised regarding waste's 'life' (or 'afterlife') as a socio-material phenomenon, including the social implications of more complex transformations that materiality undergoes in processes of recycling, reuse, and repair. Following Gille, the goal of waste research should be to raise questions as to *how* waste is produced as a socio-material phenomenon. In order to attend to the production of a socio-material order, I map the operations of what Gille (2010) calls *waste regimes* that are governing the 'production, circulation, and transformation of waste as a concrete material' (1056). The concept of waste regimes addresses the production of specific wastes through particular regimes of value and techniques of governance, including categorization, by making manifest the economic, techno-scientific, and cultural rationalities that have shaped them. Implicated in this idea of waste regimes is that the definition of waste is not a substantive or universal one, but contingent on situated and reversible relations between humans and non-humans. That is to say, waste is the product of what Barad (2003) calls *intra-action* between the discursive and the material, namely 'specific exclusionary practices embodied as specific material configurations of the world (i.e., discursive practices/(con)figurations rather than "words") and specific material phenomena (i.e., relations rather than "things")' (814). Through the productive process of intra-action, material-discursive assemblages are formed that subject matter to particular technical ordering systems and economic and cultural structures of value. Highlighting such intra-actions forms one way of understanding the emergence of an apparent order of entities: particular objects that are *bounded*, rather than amorphous, and *recognizable*, rather than uncategorized. Intra-actions that successfully consolidate a material-discursive order also express territorial control, namely the ability to monitor, regulate, and control matter and the environment.

Accordingly, this chapter treats waste as the result of positive discursive-material articulations. Yet waste as matter can be particularly elusive and the question is how waste regimes deal with this. While 'waste' forms a decisive discursive designation that is ordering matter in some contexts, it also happens to be, Moser argues, a 'category of transition, a limit category' (quoted in Gabrys 2013, 17). Transient matter defies more permanent fixture and potential for heterogeneous futures exerts itself as manifold, not-yet-actualized, trajectories of becoming. When the brand logo on an iPhone becomes detached from the device and its unity disintegrates, implicated materialities may transform into hazardous substance, a particular kind of raw material that forms a new product, or a reassembled, so-called refurbished, phone.

Processes of intra-action, according to Gabrys (2013), inform 'how and why objects hold together, and what resources are at play in both stabilizing and destabilizing those objects' (59). Waste forms an unstable assemblage of heterogeneous forces whose trajectory of becoming features underdetermined intra-actions. Yet unstable intra-actions do not signify the absence of power altogether. Rather than the vitalist material agency *pur sang*, the problematic I address pertains to the ways in which e-waste management engages e-waste's elusive, transient quality. As I argue in the latter part of this chapter, flexibility is key to 'reverse logistics.' The latter encompass the optimized recovery of reusable components and valuable materials and include waste 'all the way from used products no longer required by the user to products again usable in the market' (Li and Tee 2012). Rather than strict control over the mobilities and transformations of e-waste, and therewith over Hong Kong's borders and territory, reverse industries cultivate the logistical capacity for flexible and tactical interventions into e-waste as aleatory, global flow.

What is specific, though perhaps not unique, to Hong Kong is that waste regimes alternate between constraining and exploiting transience. As both an administrative territory and a logistical hub hosting (formal as well as informal/illegal) recycling industries, Hong Kong shows the intersection of—and contradictions between—heterogeneous waste regimes as well as the spatial registers they inscribe. Here, paradoxically, e-waste's transient nature appears both constructed and de facto emergent in an uncontrollable fashion.

E-Waste: Product or Residue

This section focuses on value recovery in the formal e-waste recycling sector. I present three situations, revolving around distinct sets of intra-actions that position e-waste as alternatively a *product*—carrying either use or exchange value—or a (hazardous) *remainder*.

In the first situation, e-waste is supposed to transform from remainder to product through reuse after redistribution to the urban poor. In Hong Kong, the local NGO Caritas combines youth employment, environmentalism, and corporate social responsibility (CSR) in an innovative programme that trains youth to recycle e-waste and prepare devices for reuse by underprivileged families in Hong Kong. The NGO receives substantial amounts of donated hardware that have been phased out, especially from big corporations in Hong Kong's knowledge industries.

Through a deal with Microsoft's CSR branch, the NGO is able to install cheap software licences on the computers. If a received device is broken, the NGO will repair it, and it will also try to recombine parts from different computers with the purpose of improving functionality and expected duration of use. For instance, the RAM can be upgraded in this way. Parts that cannot be salvaged through reuse are sorted, categorized, and sold to certified recyclers to provide a stream of revenue for the NGO. When donations are redistributed for reuse, e-waste transitions from remainder to product with use value. If recycled and sold as sorted material, e-waste transitions into a product with exchange value and a resource for a new cycle of manufacturing.

A second way in which e-waste transitions from remainder to product is through commercial recovery of materials. Waste is a potential resource of exchange value and hence an opportunity for profit. For instance, originating from Hong Kong, the recycler Li Tong Group serves over 100 electronics manufacturers around the world, such as Original Equipment Manufacturers (OEM), Original Design Manufacturers (ODM), and Electronics Manufacturing Services (EMS). These can be manufacturers such as Apple and Samsung and their subcontractors such as Foxconn. Dealing with OEMs with take-back programmes, Li Tong's website states the corporation has perfected systematic methods that 'Capture the maximum value of their redundant supply chain assets' and 'Enable an optimal rate of reuse and reapplication for parts & components.' Non-business actors equally tend to reckon on the commercial potential of waste. The manager of the e-waste branch of the St. James' Settlement jokingly confessed to me that he had previously been a 'star' seller managing an electronics store. He had made a career switch to the NGO sector in order to 'make up for sins' such as selling large numbers of televisions and what he called 'entertainment electronics.' However, he also reasoned, 'From an economic perspective, waste is just something not yet put into consideration within the market operation system. [I]f you see [items] as rubbish, they will just end up in the landfill. So, it is a problem of positioning. What I think is that there is no rubbish at all, it just depends on how you are going to use them' (Personal interview, 9 May 2014).

In a personal interview with the Environmental Protection Department (EPD) of the Hong Kong government, the bureaucrat who made time to talk to me and my research assistant argued that e-waste could be a profitable business in Hong Kong once a 'critical mass' was generated (Personal interview, 6 June 2015). In this calculation for private-sector solutions, the amount of e-waste retained in Hong Kong was perhaps even too little rather than too much. Waste could be transformed into value and the fact that it had no value initially was only conducive to the prospect of profit. Indeed, incentivizing the treatment of e-waste as a product has been the demand for raw materials in China's vast development and manufacturing industries in the 1990s and 2000s. The intake of e-waste has functioned as a channel for acquiring such materials.

As both the first and the second situations I analyze suggest, the definition of waste is not substantive but contingent on situated and reversible relations between humans and non-humans, indicating the transience of matter and its inherent potential to attain value again. The narratives accompanying logistical practices tend to imagine the possibility of what I call 'magical circuits,' in which all waste is redeemed. In other words, magical circuit narratives are undergirded by assumptions that waste will transition into something else upon its reinsertion into circulation. Yet such narratives are problematic because they fail to acknowledge the extent to which some materials remain 'unrealized either as use or as exchange value' (Gille 2010, 1054), or, as in Gabrys' view, 'Waste always returns. Even with extensive attempts to salvage, recuperate, and recycle waste, remainders surface and resurface' (2013, 132). For instance, while receiving 20,000 computers annually for recycling, Caritas has a shortage of requests for donation, and only 2000 computers are sent out. Moreover, as my Caritas interviewee argued, 'The logistical costs are really high for us and each collected donation is not necessarily 'rewarding.' […] So, the workload is quite high, that's why we have so many orders waiting because of lacking resources and work force' (Personal interview, 14 May 2015).

Material recalcitrance and ungovernability may be understood here in terms of the sheer quantity of e-waste that cannot be redeemed through redistribution schemes. Furthermore, the assertion 'waste always returns' can be taken in the sense that the extraction of valuable materials itself is bound to generate waste (Maxwell and Miller 2012, 3). For instance, recycling often generates emissions due to transportation of materials and waste water, as well as potentially air and soil pollution. As the advocacy group BAN argues, recycling is often 'a misleading characterization of many disparate practices – including demanufacturing, dismantling, shredding, burning, exporting, etc. – that is mostly unregulated and often creates additional hazards itself.'

While advocating market-oriented solutions to e-waste, my interviewee at EPD indeed acknowledged that recovery through the commercial circuit was not always possible (Personal interview, 6 June 2015). The pursuit of profit could exclude certain materialities from treatment by the private sector. Hence, he argued that businesses should take care of items containing precious metals and other recyclables. Yet public facilities ought to take responsibility for items such as cathode ray tubes that are mostly composed of less valuable plastics but contain toxic substances such as lead, barium, cadmium, and fluorescent powders. Hence, besides providing subsidies to the private sector, the government should constitute a 'fall-back' option by handling waste residues that were not sufficiently profitable. The 'natural distribution' of roles between the private and public sectors that the EPD bureaucrat invoked reflects the fact that a flawless reintegration of waste into circuits of profit and renewed production is unattainable. Moreover, it was up to the public sector and NGOs to attend to different sets of value such as social and environmental ones.

Positioned as a remainder, waste is not supposed to circulate without oversight. The United Nation's *Basel Convention on the Control of Transboundary Movements of Hazardous Wastes and their Disposal* prohibits the transfer of hazardous materials from developed to developing countries through trade or dumping. While not being a signatory to the Basel Convention directly, Hong Kong has incorporated regulation informed by the Basel Convention's ban into the Waste Disposal Ordinance in 2006. Additional regulatory frameworks were provided by a Memorandum of Understanding with mainland China in 2000, which was revised and retitled 'Cooperation Arrangement on Control of Waste Movements between the Mainland and HKSAR' in 2007. Moreover, Hong Kong has taken steps to adapt the *Waste Electrical and Electronic Equipment Recycling Program* (WEEE) directive which originated from the European Union. The city has committed to protocols for separating hazardous materials and Extended Producer Responsibility schemes that include a recycling fee and take-back programmes. Together, legal measures by the government and complementary certification by civil and industry organizations control the mobility and transformation of e-waste when it is positioned as a *remainder* that has no value (at least in its current state) or is even hazardous.

Ambiguity and/as Governance

The above section described three situations that elaborate the transition of waste into a product/resource and the separation of useless or even hazardous remainders. In practice, however, e-waste's ambiguous status as product/resource or remainder—its indeterminate nature and value as object—may complicate such governance. This section and the next investigate to what extent ambiguity implies either disorder as the failure of governance or a reconfiguration of governance.

In our personal interview, the former Greenpeace employee and coauthor of the 2003 report *Hong Kong: E-Waste Freeport?* argued, 'It is really complex. In our understanding or according to the common sense, e-waste would include electronic equipment, right? But for China, you cannot say that you cannot import e-waste [despite that they prohibit it]. Take wire. The copper of the wire is part of e-waste yet China has decided that it is okay to import it' (Personal interview, 11 June 2015). Similarly, the website of Hong Kong's EPD provides a 'comprehensive list of "prescribed hazardous waste" subject to waste import/export permit control.' Yet secondhand notebooks, despite containing such materials, can be regarded as useful devices, if not items containing valuable raw materials. They can be imported into China after payment of some special tax or after an application for permission. Yet even if secondhand devices or raw materials potentially have utility in a developing region (notably, expanding access to information and communication and meeting the needs of development for raw materials), such redemption is not guaranteed: after importation, secondhand devices can quickly breakdown or turn out to be dysfunctional in the new environment. At the same time, designations such as 'toxic' and 'hazardous,' which are seemingly objective and as clear as the distinction between 'life' and 'death,' are negotiable: toxicity is only one potential future inherent in the materiality of an electronic device, but so is functionality or value.

Moreover, e-waste's categorization shifts as it travels through multiple jurisdictions with different legislation. This can be the case even between two US states, or between Hong Kong and China, as permitted by the 'one country, two systems' principle (Lepawsky 2012; Personal interview, Hong Kong, 11 June 2015). Temporalities of regulatory change in Hong Kong and mainland China at times have been disjointed, and even though directions overall were similar, details in regulation and implementation have differed. The *E-Trash Transparency Project* (Puckett et al. 2016) by

BAN notes that at the time of publication, the Hong Kong Waste Disposal Ordinance has 'unique definitions from the Basel Convention with respect for example to circuit boards, which they do not necessarily consider hazardous' (103). This means Hong Kong accepts the inflow of central processing units (CPU), printers, faxes, keyboards, mice, yet the condition that licensed recyclers handle them is often not met.

As analyzed by Foucault (2007, 84–5), disciplinary power consists in the capacity 'to separate so the normal and abnormal are classified against an imposition or structural consistency ("the permitted and forbidden")' (Dieter 2011, 197). For e-waste, this would mean that a priori legal or moral models are applied to decisive categorization and that material processes and practices are subjected to strict oversight. For instance, the categorization of waste and designation of toxicity are unequivocal and import and export are strictly prohibited. Drawing on Barad's terminology, successful disciplinary governance of e-waste would mean decisive intra-actions that construct e-waste as a stable discursive-material assemblage.

However, as the above suggests, categorization for e-waste is flexible and contingent on territorial differences. On the ground at a shipping port, exhaustively checking the contents of containers is immensely difficult, because while some mobilities are permitted and even encouraged, it is less clear 'what is what': What is legal trade, and what is a crime? What is a product/resource, and what is a useless or hazardous remainder? Rossiter (2016, 86) argues that standardization expresses control throughout an integrated region or zone; however, efforts to standardize prompt new tactics of circumvention and instances of incommensurability, too. At an international port like Hong Kong's, the reliance on codes deriving from the international goods nomenclature system to report cargo at customs might result in nothing more than a self-referential informational order. As BAN argues in a briefing, issued on 20 January 2016, a special code for e-waste is missing from the Harmonized Tariff Schedule maintained by the World Customs Organization and, even if there was one, e-waste traders could avoid using it. There is a practice of reporting codes designated for less-surveilled legal imports, such as metal scrap, even when the cargo is of mixed composition and includes prohibited materials. Despite the deployment of X-ray scanning for containers, which is less labour-intensive than inspection without such a scopic prosthesis, the matter remains rather elusive and hard to govern (e.g., for a detailed account of Antwerp's port, see Bisschop 2016, 85–7).

Exacerbating the deficit of disciplinary power, there is a lack of a clear definition of sustainability in 'green' recycling. Examples of global certification programmes for responsible disposal are the R2 standards and the e-Stewards. Yet these standards represent rather different approaches and practices. Counting Li Tong Group among its certified businesses, R2 is partially designed by the scrap industry and treats the export of used electronics 'largely as a process of commodity circulation, "resource recovery", and "bridging the digital divide"' (Pickren 2014, 27). Certification is supposed to stimulate the sector's reform. The e-Stewards programme is administered by BAN and endorsed by major environmental groups. Its certification programme prescribes a voluntary ban on export from developed countries to listed developing countries of what they unequivocally label as hazardous waste (27). In Hong Kong, the government promotes the ISO 14001 standard for environmental management. Yet while this standard adds to the reputation of companies that are certified under it, it does not prescribe specific norms or targets related to environmental impact. Rather, ISO standards provide codes of conduct and assess procedures of management. They do so without inviting public political dialogue and are not even publically available (Easterling 2014). Confounding standards for certified recycling exacerbate the lack of clarity with regard to corporate claims to being 'green.' As Maxwell and Miller (2012) argue, 'sustainability' often refers to a cost-benefit analysis that seeks a balance between the demands for profits on the one hand and the costs of ecological care on the other. This calculation may fail to adequately take into consideration the non-negotiable limitations of planetary resources and hence does not result in practices that are truly sustainable from an ecological point of view.

The lack of data regarding the afterlife of disposed goods further impairs disciplinary oversight. A chart from EPD, included in the Friends of the Earth report (2011), shows that in Hong Kong, in 2009, 97.7 per cent of e-waste that was recycled (rather than dumped in local landfills) passed through unknown channels and a meagre 2.3 per cent passed through channels supervised by the EPD. While these numbers do not represent the size of legal versus illegal waste flows, they underscore the lack of oversight and control over waste flows. Critiquing the situation, my interviewee at Friends of the Earth argued, 'They [EPD] only get the numbers and statistics from the people who report to the department. But whether those are true or not actually no one knows, because Hong Kong is a free market. So they don't actually have the real numbers' (Personal interview, 2 September

2015). Meanwhile, Greenpeace has estimated that salvage of e-waste through the informal and illegal circuits comprises 75 per cent, meaning that this portion of collective e-waste emission has simply 'disappeared' (Maxwell and Miller 2012, 114). The United Nations Environmental Program (UNEP 2015, 7) declares that globally the amount of e-waste that is recycled and disposed of properly ranges 'between 10 to 40 percent according to different estimates,' meaning that 60–90 per cent is either informally/illegally traded or simply dumped. The wide margin cited here is a symptom of the very problem the percentiles address, namely, the deficit of disciplinary oversight over e-waste flows.

FLEXIBLE LOGISTICS AND ALEATORY FORCES

However, the semantic and legal ambiguities rendering the category of e-waste unstable do not necessarily signal the failure of e-waste management. In discussing the emergence of power as security (rather than discipline), Foucault argues that walls around cities became obsolescent in the eighteenth century, upon which the task for governance was formulated no longer as protecting and controlling a territory but as distinguishing between 'good' and 'bad' circulation. Security consists furthermore in technologies of power prospecting potential threats and imminent futures that are 'not exactly controllable, not precisely measured or measurable' (Foucault 2007, 35). By analogy in e-waste governance, it is about managing mobilities rather than inhibiting them. Major stakeholders across the public, private, and NGO sectors embrace circulation as the answer to the waste problem with the notable exception of BAN, which rejects the export of waste to designated developing countries. Discussing labour migration flows, Mezzadra and Neilson (2012) suggest that instead of the border as a separating line between discrete spaces, multiple techniques of bordering produce spaces and organize flows. Rather than simply effectuating a rule of inclusion/exclusion, these techniques and measures enable partitioning, filtering, and hierarchization of flows. At stake are 'multiple parameters that in combination determine the vectors of movement' (69–70). Likewise, in the management of e-waste, special permits and fees that enable import/export render borders multiple and porous rather than singular and rigid, as discussed above. In this process, the risk of flows disappearing from the monitoring systems is accepted. Even so, it deserves to be mentioned that EPD, in response to BAN's *E-Trash Transparency Project*, issued a press release on 17 June 2016 stating that

'The EPD will not tolerate any hazardous e-waste being illegally imported to Hong Kong.' Moreover, in our interview, the EPD representative admitted to concerns over the exportation of e-waste to developing areas in Asia. Yet he also maintained that e-waste's treatment as 'business as usual' implies that the government will not track and oversee everything that is handled by the private sector (Personal interview, 6 June 2015). While this attitude exemplifies a widespread neoliberal governance style (Maxwell and Miller 2012, 117), the strategic performance of non-governance may function here as a resource for governance (Rossiter 2009, 2016, 79). Non-governance helps the 'ICT Hub' and its knowledge economy rid itself of infrastructures and gadgets that all too quickly obsolesce due to the sped-up cycles of innovation the city adheres to.

Perhaps the point needs to be pushed even further. After all, it is hard to view the lack of oversight and accurate data—together with the estimated proportion of e-waste that disappears globally to end up in informal and illegal circuits or is dumped (75 per cent according to Greenpeace and 60–90 per cent according to UNEP)—as merely the accidental slippage concomitant with 'proper' flow, as the security paradigm would have it. For governance actors such as SERI, which administers the R2 standard, whereas risk is permissible, lending e-waste the 'freedom' to transform into something else generates *opportunity*. Actualization of e-waste's potential for value and profit is left to free enterprise—namely corporations and entrepreneurs chasing opportunities the world over and pioneering innovative recycling technologies—to turn threat into profit and productivity. The question is to what extent this quest for opportunity in a condition of normalized crisis displaces security as the management of mobilities. Massumi's (2009, 157) argument might hold sway that an emergent modality of power—distinct from security—takes its environment as riddled with 'anywhere-anytime potential for the proliferation of the abnormal,' and uncontrollable, aleatory forces. This permanently critical condition of instability and lack of normality is enabling, though, as it provides unforeseen opportunities.

This form of power, however, does not work by either discipline or security but, paradoxically, by 'fleeing forward': 'It lives out its instability. It is emergent order on the edge, riding the wave-crest of everywhere-apparent chaos' (176). Massumi's conceptualization of environmental power allows us to understand the attenuated control over the borders of Hong Kong and its territory as not the end of power altogether but a transformation of its modality. Moreover, it allows us to come to terms with the fact that a 'developed' city and supposed 'ICT Hub' can become a haven for e-waste trade and even a site where material accumulates and lingers. While Hong Kong's advanced recycling and reverse logistics industries engage e-waste as aleatory flow, the city also faces a loss of control over territory and borders, which manifests itself in ecological harm, as activists warn. Consequentially, e-waste mobilities and transformations witnessed in Hong Kong are undermining the spatial registers delineating 'territories' and 'regions' and designating them as either 'developed' or 'developing.'

To corroborate this assessment of the operation of power in the context of e-waste, it is useful to investigate the ways in which logistical practices and their technologies seek opportunity in e-waste's mobility and transformation. The key is that e-waste is not constructed as an object to be governed but as an uncharted field of possibilities and potential to be exploited. Logistical software as well as the interfacing between formal and informal circuits play a role in garnering the capacity to tackle and exploit this potential. Yet, as I will argue, by substituting for a disciplinary form of governance that seeks control over e-waste's mobility and transformation, logistical management also *produces* e-waste as elusive and transient. Governance cannot and does not control the ramifications that emerge in practice.

Following technical discourses in reverse logistics, e-waste lacks a singular character and forms a field of possibilities instead. Whereas its transformative potential for manifold futures renders e-waste a rich resource, actually exploiting it requires calculating and forecasting the interplay of the many shifting factors at stake in value recovery. Gabrys (2013, 59) argues that waste mimics the post-industrial commodity that comes into being through just-in-time production, characterized by flexible, adaptive processes that integrate market information feedback. More so, experts contend that reverse logistics tackle degrees of complexity and uncertainty even higher than those present in supply chain management due to the many actors, processes, and unpredictable push-and-pull factors implicated. Academic literature in fields such as management claims to transcend 'traditional optimisation approaches' by accounting for the 'dynamic behaviour of closed-loop supply chains' and 'the high complexity and connectivity inherited with various value recovery processes' (Lehr et al. 2013, 4105). On the one hand, the literature prescribes methods to tackle factors of uncertainty specific to reverse logistics by means of intensified informatization and integrated data usage throughout the reverse and supply chains, including by partners. On the other hand, the literature presents probabilistic models and approximation techniques in order to optimize value while incorporating uncertainty into the calculation (Fleischmann et al. 2003; Kokkinaki et al. 2003; Li and Tee 2012). For instance, for OEMs with take-back programmes, optimizing the reverse chain happens by means of logistical software that calculates to what extent costs would be saved by recovering parts for reuse in manufacturing or, alternatively, by turning e-waste into raw materials after accounting for the prognosticated needs of its supply chain. Yet another option is sales of raw materials in external markets whereby real-time information on quantities at the supply side as well as market prices for sales is taken into account. Reverse logistics software can further help determine the most strategic locality for particular recovery activities, taking into account multiple factors including tax costs pertaining to importation and potential costs associated with legal liabilities as well as transportation and labour costs at specific moments.

The assertion that reverse logistics face extraordinary complexity and uncertainty is concomitant with an emphasis on the import of flexibility, agility, and adaptability to exploit unforeseen opportunities. Developers of specialized reverse logistical software promote such programmes as unmissable tools to garner the capacity for adaptive, flexible, tactical, and inventive operation that is necessary to tackle and exploit e-waste. Hence, Oracle talks of 'agile and effective' logistics, while UPS promises that vigilant reverse logistics uncover 'unexpected' revenue sources and 'hidden profits' while 'taking advantage of every opportunity to squeeze more revenue from all returned products' (Oracle 2014; Greve and Davis 2012). Similarly, a report by PricewaterhouseCoopers contends that reverse logistical chains that are enhanced by the 'virtual corporation' are able to 'exploit profitable opportunities in a volatile market.' The report adds that 'agility is needed in less predictable environments where demand is volatile and the requirement for flexibility is high' (Rosier and Janzen 2008, 70).

The advocated tactical, flexible, and adaptive techniques of e-waste management do not seek to control but to exploit an environment that is understood to be highly complex, dynamic, and uncertain. From a disciplinary perspective, the ensuing condition denotes disorder and a failure of governance due to the lack of control over the mobility and transformation of e-waste. After all, discipline operates on and through identifiable, categorical objects, but it is undermined by elusive, transient, and aleatory forces. Yet what Massumi describes as environmental power feeds on e-waste as an uncharted field with possibilities and transformative potential by harnessing techniques and capabilities to tackle and exploit the unpredictable, disruptive, and unforeseen.

Following Massumi, the aleatory, transient quality of e-waste is both constructed and produced in governance (or, governance as non-governance) *and* emergent in practice in ways uncontrolled by governance. That is to say, the 'unforeseen' might exceed what gets tackled and what governance *can* control. This is the case when e-waste dissipates into the informal circuit. It should be noted, however, that the latter is not marginal but vast and pervasive. Informal circuits interface with formal circuits at many points, maintaining either a symbiotic or anti-thetical relationship (Bisschop 2012; UNEP 2015, 34). Informality offers specific 'advantages' to logistical processes that benefit from flexibility and ad hoc operation. For instance, its flexibility in collecting e-waste from all the corners of the city and facilitating price negotiations with individual owners on the spot remains unmatched by the formal circuit. In Hong Kong, informal collectors go door to door or chalk contact numbers on makeshift boards placed at street corners. Their work continues to undermine the functioning of municipal collection points and take-back programmes. Other facets of the flexibility afforded by informality are grimmer. For instance, labour is employed on extremely precarious terms, hired per day as the need arises, or self-employed. In Hong Kong, undocumented migrants and asylum seekers who lack work permits often end up in e-waste yards. Impoverished elderly women are generally the ones who wait all day long on street corners to collect disposals from passers-by. Furthermore, facilities can be opened and closed down rapidly, without compliance with environmental regulations and other laws. Anonymity is a resource for informal businesses (complementing non-governance as a resource for administrations) that enables legal risk-taking when processing or dumping toxic and unprofitable components: the absence of a corporate reputation to be guarded lowers the stakes.

If, on the one hand, governance (as non-governance) and logistical management construct e-waste as an aleatory, transient force that is rendered productive through its 'free' circulation, on the other, the ramifications emerging in practice are truly beyond control and ungovernable. One such ramification is the massive counterfeit production as well as a burgeoning DIY (Do-It-Yourself) culture that is resourced by the informal/illegal circuit at local markets such as Chung King Mansion, Sin Tat, and Sham Shui Po in Hong Kong, and even more so Huaqiangbei in Shenzhen.

Cross-border informal networks extending into mainland China prefigured the recent trend in formal industries to integrate recycling with manufacturing by feeding recovered e-waste into the supply chain. Yet the recovery activities of the informal circuit have undermined the intellectual property interests of the formal OEMs as well as the hegemony of the ICT Hub's knowledge economy (Cf. Hu 2008; Pang 2012, 222).

More alarming than such violations, to me at least, are the environmental ramifications that are beyond control. We hear of toxins seeping into the ground in the New Territories in Hong Kong. Piles of e-waste lay abandoned on rural lands, long after informal recycling facilities have shut down. Vapours and fumes circulate in the air when an accidental fire breaks out, which has even caused a Shenzhen newspaper to turn the tables and accuse Hong Kong of polluting Shenzhen's air. In these cases, non-disciplinary governance fails to prevent e-waste-related accidents that no longer can be considered truly accidental in the sense of occurring 'by chance.'

CONCLUSION

As I have argued in this chapter, effective disciplinary governance of e-waste is complicated by the ambiguous positioning of waste as remainder on the one hand and product or resource on the other. The transient character of waste matter and the manifold potential for trajectories of becoming inherent in it inhibit strict categorization and oversight. Moreover, legal differences and the semantic ambiguity pertaining to 'sustainability' imply a move away from unequivocal norms separating the legal and the illegal, or the harmless and the toxic. Yet power may not have faded in the face of ambiguity, risk, and non-governance. Rather, non-governance may be integral to governance in the so-called ICT Hub, which is bound to generate an excess of e-waste. Nonetheless, the relation between the formal knowledge economy of the ICT Hub and the informal zones of e-waste recycling that traverse Hong Kong's territory is ambivalent and often conflictual, as with regard to intellectual property.

Rather than casting waste as essentially 'disorder' and the opposite of social or human endeavour, as critiques of capitalism and vitalist-materialist approaches both tend to do, I suggest consideration of waste's involvement in binaries of order/disorder in relation to diverse modalities of power. First, disciplinary power deploys strict definitions of waste, predicated on toxicity and harm, and expresses itself as control over waste matter and its mobility as well as the territory. In contrast, security manages 'good' and 'bad' circulation in order to maintain some kind of liveable equilibrium: a somehow 'sustainable' world in which we evade (direct) ecological disaster that endangers population health, yet at the same time there is minimal interference in markets. Third, environmental power capitalizes on waste as a global, aleatory force, exploiting waste's manifold potential as well as semantic and legal ambiguities and surreptitiousness. Disciplinary power, security, and environmental power respectively articulate moral and legalistic, biopolitical-governmental, and entrepreneurial attitudes. Yet it would be a mistake to ascribe these attitudes exclusively to any particular actors such as NGOs, states, or corporations.

Examination of logistical practices and their technologies can bring us closer to an understanding of the multiple modalities of power pertinent to e-waste, as well as the conflicts and contradictions they provoke in situ, since the governance techniques and objectives that inform them are incongruent. In ancient Greek, *logistikos* shares its etymological root with *logos*, or rationality, and refers to the skill of calculation of supplies needed for the battlefield in the context of war (Cowen 2014). However, in order to have a more nuanced grasp of the ways in which logistical practices entail modalities of power, our association of logistics with a singular *logos*—a technocratic rationality that is optimally efficient and globally homogeneous—needs revision. The rise of specialized software for reverse logistics, which is tasked with discovering unforeseen profits, as well as the massive operation of the informal/illegal circuit in e-waste recycling, underscores the import of flexibility in logistics. The capacity for ad hoc, adaptive, and inventive operation supports the exploitation of e-waste as a resource, but also constructs it as an elusive, aleatory, transient force. This approach undermines disciplinary control over the mobility and transformation of e-waste. The ramifications are 'accidents' (no longer of a truly accidental nature) that governance cannot and does not prevent. My analysis referred to the counterfeit production of electronics as well as precarious, unregulated labour, and environmental harm. By exploring logistical practices and their technologies, we can start to conceive of the astonishing ratios of e-waste that go off the radar, not in terms of accidentalness and disorder, but in terms of shifting modalities of power.

REFERENCES

Barad, Karen. 2003. 'Posthumanist Performativity: Toward an Understanding of How Matter Comes to Matter.' *Signs: Journal of Women in Culture and Society* 28, no. 3: 801–31.

Bennett, Jane. 2010. *Vibrant Matter: A Political Ecology of the Thing*. Durham: Duke University Press.

Bisschop, Lieselot. 2016. *Governance of the Illegal Trade in E-Waste and Tropical Timber: Case Studies*. London: Routledge.

Bisschop, Lieselot. 2012. 'Is It All Going to Waste? Illegal Transports of E-Waste in a European Trade Hub.' *Crime, Law, and Social Change* 58: 221–49.

Office of the Government Chief Information Officer. 2017. *Hong Kong: A Premier Location as an ICT Hub*. Accessed 13 October 2017. https://www.ogcio.gov.hk/en/facts/doc/Fact_Sheet-HK_as_ICT_Hub-EN.pdf

Chung, Shan Shan, Ka-yan Lau and Chan Zhang. 2011. 'Generation of and Control Measures for E-Waste in Hong Kong.' *Waste Management* 3: 544–54.

Cowen, Deborah. 2014. *The Deadly Life of Logistics: Mapping Violence in Global Trade*. Minneapolis: University of Minnesota Press.

Dieter, Michael. 2011. 'The Becoming Environmental of Power: Tactical Media after Control.' *Fibreculture Journal* 18: 177–205.

Easterling, Keller. 2014. *Extrastatecraft: The Power of Infrastructure Space*. London: Verso.

Fleischmann, Moritz, Jacqueline M. Bloemhof-Ruwaard, Patrick Buellens, and Rommert Dekker. 2003. 'Reverse Logistics Network Design.' In *Reverse Logistics: Quantitative Models for a Close-Loop Supply Chains*, edited by Rommert Dekker, 65–94. Berlin: Springer.

Foucault, Michel. 2007. *Security, Territory, Population*. New York: Palgrave Macmillan.

Friends of the Earth. 2011. *The Policy Study Report on the Waste Electrical and Electronic Equipment Directive*. Accessed 14 October 2016. https://www.foe.co.uk/.../report-influence-eu-policies-environment-9392

Gabrys, Jennifer. 2013. *Digital Rubbish: A Natural History of Electronics*. Michigan: University of Michigan Press.

Gille, Zsuzsa. 2010. 'Actor Networks, Modes of Production, and Waste Regimes: Reassembling the Macro-Social.' *Environment and Planning A* 42, no. 5: 1049–64.

Greenpeace. 2003. 香港:電子毒物自由港? *[Hong Kong: E-waste Freeport?]*. Accessed 15 October 2016. http://www.greenpeace.org/hk/press/releases/toxics/2003/08/96293/

Greve, Curtis, and Jerry Davis. 2012. 'Recovering Lost Profits by Improving Reverse Logistics.' Accessed 15 October 2016. https://www.ups.com/medi/en/Reverse_Logistics_wp.pdf

Hu, Kelly. 2008. 'Made in China: The Cultural Logic of OEMs and the Manufacture of Low-Cost Technology.' *Inter-Asia Cultural Studies* 9, no. 1: 27–46.

Jackson, Steven. 2014. 'Rethinking Repair.' In *Media Technologies – Essays on Communication, Materiality and Society*, edited by Tarleton Gillespie, Pablo Boczkowski, and Kristen Foot. Cambridge, 221–39. Cambridge, MA: MIT Press.

Kokkinaki, Angelica, Rob Zuidwijk, Jo van Nunen, and Rommert Dekker. 2003. 'ICT Enabling Reverse Logistics.' In *Reverse Logistics: Quantitative Models for a Close-Loop Supply Chains*, edited by Rommert Dekker, Moritz Fleischmann, Karl Inderfurth, and Luk N. van Wassenhove, 381–406. Berlin: Springer.

Lehr, Christian, Jörn-Hendrik Thunb, and Peter Milling. 2013. 'From Waste to Value – A System Dynamics Model for Strategic Decision-Making in Closed-Loop Supply Chains.' *International Journal of Production* 51, no. 13: 4105–16.

Lepawsky, Josh. 2015. 'The Changing Geography of Global Trade in Electronic Discards: Time to Rethink the E-Waste Problem.' *The Geographical Journal* 181, no. 2: 147–59.

Lepawsky, Josh. 2012. 'Legal Geographies of E-Waste Legislation in Canada and the US: Jurisdiction, Responsibility and the Taboo of Production.' *Geoforum* 43, no. 6: 1194–1206.

Li, Richard and Tarin Tee. 2012. 'A Reverse Logistics Model for Recovery Options of Ewaste Considering the Integration of the Formal and Informal Waste Sectors.' *Procedia: Social and Behavioral Sciences* 40: 788–816.

Massumi, Brian. 2009. 'National Enterprise Emergency: Steps Toward an Ecology of Powers.' *Theory, Culture & Society* 26, no. 6: 153–85.

Maxwell, Richard and Toby Miller. 2012. *Greening the Media*. Oxford: Oxford University Press.

Mezzadra, Sandro and Brett Neilson. 2012. 'Between Inclusion and Exclusion: On the Topology of Global Space and Borders.' *Theory, Culture & Society* 29, nos. 4–5: 58–75.

Moore, Jason. 2015. *Capitalism in the Web of Life: Ecology and the Accumulation of Capital*. London: Verso.

Oracle. 2014. 'Oracle's Depot Repair Implementation: Driving Product Take-Back and Recycling at Oracle.' White Paper. Accessed 15 October 2016. www.oracle.com/us/products/applications/green/oracle-reverse-2320318.pdf

Pang, Laikwan. 2012. *Creativity and Its Discontents: China's Creative Industries and Intellectual Property Rights Offenses*. Durham and London: Duke University Press.

Parikka, Jussi. 2015. *A Geology of Media*. Minneapolis: University of Minnesota Press.

Puckett, James, Eric Hopson, Eric Huang, and Monica Huang. 2016. *Disconnect: Goodwill and Dell, Exporting the Public's E-Waste to Developing Countries*. Basel

Action Network (BAN). Accessed 21 March 2017. http://www.ban.org/news/2016/5/9/goodwill-and-dell-inc-exposed-as-exporters-of-us-publics-toxic-electronic-waste-to-developing-countries

Rosier, Mathieu and Bertjan Janzen. 2008. 'Reverse Logistics. PricewaterhouseCoopers.' Accessed 15 October 2016. www.pwc.nl/nl/assets/documents/pwc-reverse-logistics.pdf

Pickren, Graham. 2014. 'Political Ecologies of Electronic Waste: Uncertainty and Legitimacy in the Governance of E-Waste Geographies.' *Environment and Planning A* 46, no. 1: 26–45.

Rossiter, Ned. 2016. *Software, Infrastructure, Labor: A Media Theory of Logistical Nightmares.* New York: Routledge.

Rossiter, Ned. 2009. 'Translating the Indifference of Communication: Electronic Waste, Migrant Labour and the Informational Sovereignty of Logistics in China.' *International Review of Information Ethics* 11 (October): 35–44.

United Nations Environmental Program (2015) *Waste Crime – Waste Risks: Gaps in Meeting the Global Waste Challenge.* Accessed 15 October 2016. www.unep.org/delc/Portals/119/publications/rra-wastecrime.pdf

CHAPTER 11

Geopolitics of the Belt and Road: Space, State, and Capital in China and Pakistan

Majed Akhter

China is repositioning itself as a major driver of global developmental and investment activity. As Donald Trump was leading the US towards economic protectionism and geopolitical isolationism, in January 2017, Chinese President Xi committed his country to 'global free trade and investment' at the World Economic Forum in Davos. Xi also urged global elites to 'redouble efforts to develop global connectivity to enable all countries to achieve inter-connected growth and share prosperity.' Large physical infrastructures, especially those geared towards increasing inter-regional connectivity, are key to this vision. This chapter develops a structural and theoretical approach to the analysis of Chinese-led infrastructure globalization.

The Chinese state has already embarked on an ambitious plan of coordinated infrastructure investments to lay the groundwork for its Asia-centric vision of globalization. Institutions such as the Asian Infrastructure Investment Bank support China's drive to increase investments in an array of large infrastructure projects. The so-called Belt and Road plan, also sometimes called the Belt and Road Initiative (BRI), will finance transportation, communications, and industrial and energy projects across Asia. Its land-based 'belt' will extend from Western China through Central Asia to Europe and the Mediterranean, and the ocean-based 'road' will connect

---

M. Akhter (✉)
Department of Geography, King's College London, London, UK

© The Author(s) 2018
B. Neilson et al. (eds.), *Logistical Asia*,
https://doi.org/10.1007/978-981-10-8333-4_11

South China to the Indian Ocean via South-East Asia. Chinese officialdom has recently offered some clarification of its BRI vision, and BRI projects will have access to a range of new sources of finance. In March 2015, a joint statement—by the National Development and Reform Commission, the Ministry of Foreign Affairs, and the Ministry of Commerce—announced that the main purpose of the BRI is to create 'strategic propellers for hinterland development.' Officials hope the BRI will catalyze continent-wide development. It aims to do so by encouraging 'countries along the Belt and Road [to] improve the connectivity of their infrastructure construction plans and technical standard systems, jointly push forward the construction of international trunk passageways, and form an infrastructure network connecting all sub-regions in Asia' (National Development and Reform Commission et al. 2015).

The political commentary on BRI tends to understand it either as a case of 'win-win' inter-state cooperation (Liu and Dunford 2016), or as China asserting its economic power to trap its neighbours in a web of economic dependency (Miller 2017). Although these 'cooperation' and 'domination' narratives of China-Asia relations seem to be diametrically opposite, their approach to geopolitical analysis is very similar. The crucial questions in this mode of geopolitical analysis, which is fixated on the state, concern the personalities, values, and long-range strategic visions of key elite state actors. Moreover, these understandings, which conflate social space with state space, fall into an analytical 'territorial trap' whereby 'China' is assumed to be an internally homogeneous space that acts on other state spaces that bound socially homogeneous spaces (Agnew 1994).

This chapter follows Samaddar (2015) to develop a more structural—rather than interest-based—approach to the geopolitics of contemporary capitalism. It does so by making two major departures from the analysis of the BRI as an instance of either cooperation or domination. The first major departure is to understand the contradictions of capital accumulation on multiple scales as a geopolitical force. Because state power—and, by extension, inter-state coordination and cooperation—is necessary for the establishment and maintenance of capitalist social and physical infrastructures, Marxist state theory approaches accumulation as inescapably geopolitical (Patnaik and Patnaik 2016; Desai 2013; Glassman 2011). I draw on the Marxist geographical theory of the 'spatial fix' to understand the political economic constraints and pressures under which states formulate their infrastructural and investment policies. I argue that the inherently contradictory nature of capitalist accumulation tends to invoke certain spatial responses from state authorities that have decisively shaped the geography of Chinese state-controlled infrastructure investments. The second major departure this chapter makes is to understand social space as something state power attempts to shape and homogenize, but which it can, in fact, never completely control (Scott 1998).

Social space is far too complex for any state or capitalist to control; it is heterogeneous, fractured, and is produced by a range of historical forces that accrue and influence over time. Social space is also increasingly produced and reconfigured through large infrastructures and the various logistical operations of globalizing capital (Mezzadra and Neilson 2013; Neilson 2012). An 'operation' of capital is 'a moment connection and capture that exhibits the materiality of even the most ethereal forms of capital' (Mezzadra and Neilson 2015, 5). Operations occur not in the imaginary smooth space created by state power but instead in the existing complex, fractured, and scaled social spaces over which states claim formal sovereignty. By paying attention to the geopolitics of accumulation and heterogeneous space, this chapter sketches a theoretical orientation to the geopolitical analysis of 'global China' and China-in-Asia.

The empirical focus of the engagement is the China-Pakistan Economic Corridor (CPEC). Chinese state officials consider the CPEC to be 'a flagship program of China's Belt and Road Initiative.' State documents imagine a '1 + 4 cooperation structure with the CPEC at the center' articulated with 'the Gwadar Port, transport infrastructure, energy and industrial cooperation' that will result in a 'boom [for] the country's economy' (Xinhua 2015). The rosy discourse from the Chinese state mirrors optimistic projections from Pakistani officialdom. During Xi Jinping's visit to Pakistan in April, Nawaz Sharif poetically urged everyone to remember that the Pakistan-China friendship was 'higher than mountains, deeper than the oceans, sweeter than honey, and stronger than steel' (Stevens 2015), and the Pakistani Minister for Planning, Reforms, and Development has repeatedly echoed the Chinese infrastructural vision of an integrated Eurasia. In December 2015, he applauded the fact that China was making Pakistan the geographical lynchpin of its continental vision due to its location at the intersection of three Asian regions of immense growth potential—South Asia, Central Asia, and China (Board of Investment 2015b). Financial investors similarly echo the optimism of central state officials. The Hong Kong-based Fung Business Intelligence issued a report on the business potential of the new Silk Road Project in May 2015 and gushed that 'The CPEC is expected to spur investments, boost bilateral trade flows between the two countries, improve infrastructure, and help ease the energy shortage in Pakistan' (Chin et al. 2015). This chapter challenges these infrastructural visions—of smooth connectivity, integration, and spaces primed for logistical efficiency—by situating the BRI and the CPEC within a larger frame that accounts for the geopolitical implications and effects of capital accumulation, and the contradictions and antagonisms of historically created social space.

The next section, 'Securing the Spatial Fix,' reviews and summarizes the theory of the spatial fix and elaborates on the special role of large physical infrastructures within this theory. It also insists on the necessity of supplementing the theory of the spatial fix with a more nuanced understanding of political—as opposed to exclusively economic—drivers and effects of the geography of capital investments. In the section 'The Spatiality of Accumulation and the China-Pakistan Economic Corridor,' a closer analysis of the spatial structures of over-accumulation in the Chinese economy and the social spaces implicated in the CPEC is presented. The aim is to illustrate principles of a more structural approach to the politics of the BRI, even while grounding the analysis in historical and geographical particularity. I conclude the chapter by summarizing its main arguments and with a suggestion for more theorized comparative analysis of globalization and the politics of large infrastructures.

Securing the Spatial Fix

I begin with the premise that social space is fragmentary and heterogeneous, a position held by many theorists of globalization (Mezzadra and Neilson 2013; Jessop et al. 2008; Harvey 2001). Social space is also uneven with respect to the distribution of economic and political power and, while capitalist accumulation is necessarily a spatial and socially uneven process, some regions and classes are becoming much richer, and doing so much faster, than others (Smith 2010; Harvey 2006). Economic factors leading to this spatial unevenness of the accumulation process include increasing returns to scale, the 'stickiness' of fixed capital investments, and specialized labour forces. Over time, as wealthy regions invest in place-based infrastructure, they not only attract greater opportunities but also become subject to constraints. That is to say, the physical infrastructures of a region become a defining part of the landscape and of the relations of power between regions in the capitalist world system (Harvey 2017). This approach contextualizes the politics of large-scale infrastructure projects within the contradictions of capitalist accumulation on a world scale.

The theory of spatial fix, as developed by the geographer David Harvey (2001, 2003), is concerned with interpreting history through an examination of contradictions within the accumulation process and the spatial strategies (such as outsourcing or just-in-time production) pursued by capitalists in response to these contradictions. Harvey's spatial fix theory allows us to understand region formation as the articulation of two types of spaces—the contiguous space of territoriality and the non-contiguous space of flows. 'Fix' has a dual meaning here. It means to affix or mobilize, or to keep fixed in the general circulation of value in the capitalist economy. It also alludes to 'fix' in the sense of solving (or at least deferring) the contradictions of capitalism—such as over-accumulation of surplus capital, underconsumption of goods by consumers, and the underprovision of necessary environmental and social services (Harvey 2006). In this broader sense, the concept of the spatial fix signals all the 'way[s] to soak up capital by transforming the geography of capitalism,' including, but not limited to, the relocation of investment or 'outsourcing' (Schoenberger 2004, 428). In other words, by taking both aspects of 'fix' into account, Harvey's concept can be usefully summarized as the elucidation of how 'socio-spatial fixes provide avenues for reproducing conditions of capital accumulation and attenuation crisis tendencies via "fixing" capital in particular territorial configurations that transform the intensive and extensive spatio-temporal rhythms of accumulation' (Ekers and Prudham 2015, 2439).

Spatial fixes often occur through the construction of large-scale infrastructures (Harvey 2001, 2003), and such infrastructural spatial fixes highlight the complex nature of contradiction in the capitalist accumulation process (Desai and Loftus 2013). While large-scale fixed infrastructures have a long gestation period and may be an effective way to soak up surplus capital during a time of over-accumulation crises, they can also create inertia for capital. For example, such large investments in embedded socio-technical systems may run the risk of being made redundant and uncompetitive, and drastically devalued, by innovations.

Fixed regional production complexes offer advantages to capitalists. It can also, on the other hand, enable regional labour to organize to the point that they are able to make effective demands on capital. Fixed infrastructures, thus, may simultaneously enable and disable, or potentially disrupt, the geographic expansion of capitalist accumulation. While fixed investments are necessary to the functioning of capitalism and indeed provide a strategy to displace the contradiction of over-accumulation that is systemic to capitalism, they are also a frustration and can act as a brake on the mobility of capital when they lead to the establishment of strongly organized communities of resistance and entitlement.

Labour geographers and historians have criticized the theory of the spatial fix for its erasure of worker agency (Cowie 2001; Herod 1997). Rather than focus exclusively on the spatial fix as a response to crises of overproduction, labour scholars have argued that worker militancy presents an equally important interruption to capitalist accumulation. This labour-oriented critique points towards larger weaknesses of the theory of the spatial fix—its lack of attention to political forces and ideologies in general. While the economic contradictions of over-accumulation generate tendencies to shift capital around and to sink surplus capital into large infrastructure, the abstract theory does not help our understanding of what happens when capital 'touches the ground' and interacts with a range of political and cultural histories and values. In the case of its concrete operations, this is where 'logistical power intersects' with 'the heterogeneity of global space and time,' that is, with specific place-based histories of state/society formation (Neilson 2012, 330). This suggests, therefore, a need to add an analysis of political and social forces and relations to the geography of capitalism contained in the theory of the spatial fix. This chapter attempts to do this by examining how the state *secures* the infrastructural operations of capital which are themselves manifestations of deeper structural contradictions within capitalist accumulation.

That is to say, this chapter proceeds from the theoretical premise that large infrastructures are not only projects of accumulation, but also of state territorialization. Power based in capital intersects with other forms of power rooted in other histories of state and social formation. This means that connective infrastructures, which form the foundation of any trans-regional logistical operation, must traverse space already made heterogeneous by uneven histories of development and state intervention. In situations where the infrastructural investments of the past have created divisions and inequalities in the social landscapes, the social space which capital must navigate is marked by antagonism. Often, the only way that a 'corridor' or 'enclave' of logistical space can be secured within such a fractured space is through militarization and securitization—a contemporary version of what Marx famously described as 'primitive accumulation.' This militarization of space may be unintended and unplanned but is nevertheless structurally necessary from the perspective of the process of capitalist accumulation. It can be recognized as a reaction to the operations of capital upon a terrain of irreducible and resilient social complexity, heterogeneity and, in some cases, antagonism. I term this militarization of fractured space to ensure the logistical operations of capital as a strategy to 'secure the spatial fix.' The next section elaborates on this approach to geopolitical analysis, which examines capital and heterogeneous social space as intersecting and exceeding state power, in the context of the proposed BRI investment programme of which the China-Pakistan Economic Corridor is a part.

The Spatiality of Accumulation and the China-Pakistan Economic Corridor

The developmental and geopolitical effects of Chinese infrastructural investments in the global South have attracted scholarly analysis and critique (Gallagher 2016; Chari 2015; Lee 2014; Brautigam 2009). Although China-Pakistan relations goes back to the 1950s, the scale of recent investment commitments suggests a turning point in the economic and strategic relationship between the two states. During a state visit to Pakistan in April 2015, Chinese Premier Xi Jinping committed US$46 billion—about 20 per cent of Pakistan's annual GDP—to fund energy and infrastructure projects in Pakistan over a decade. The China-Pakistan Economic Corridor (CPEC) is 'closely related to the Belt and Road Initiative (BRI), and therefore requires closer cooperation and greater progress,' according to the official BRI policy statement (National Development and Reform Commission et al. 2015). Thus, while the thrust of the BRI is on establishing connections to Europe via Central Asia, investment in South Asia, and especially Pakistan, also articulates with the Chinese vision for an integrated Asia. Indeed, the first funded project of the BRI is in Pakistan—the Karot Dam, on the Jhelum tributary of the Indus. However, I situate China-Pakistan relations within a structural frame of analysis that can account for both the contradictions of accumulation and the heterogeneity of social space. To this end, this section reviews structures of over-accumulation in China and China's policy attempt to direct these spatial fixes through the 'go west' and 'go out' policies. It then examines how the forces of territorialization and accumulation come together in a specific national context—the CPEC.

The Spatiality of Over-Accumulation

Contrary to liberal mythology, the capitalist market economy—in China and elsewhere—did not come to fruition when the state allowed 'free markets' to operate. Historically speaking, states have been actively engaged in nurturing territorial economies with a host of fiscal tools including taxes, currency manipulations, subsidies, standards, tariffs, customs, and duties (Flint and Taylor 2007). Similarly, minimum wage laws and social security/welfare policies enforced by the state have helped to maintain effective levels of local demand. Aside from these actions, as managers of a geographically defined economic space, state intervention is always also necessary to maintain functioning markets, construct and maintain physical infrastructure, enforce contracts, establish and protect property rights, create suitable conditions for waged labour, and enable mass consumption to take hold (Neocleous 2014). The many economic functions of state power remind us that political and economic power do not operate as abstract 'logics,' but rather are articulated in actual and complex historical situations. That is to say, even in China, Chinese state power has formed in dialectical relation to the contradictory geographies of accumulation that the state itself has helped cultivate.

China's economy has grown rapidly since the early 1980s. For almost three decades after the Communist Revolution in 1949, Chinese developmental strategy revolved around a transfer of rural surpluses to urban and industrial regions through migration controls and geographically differential inflation and interest rates. The explicit policy was thus one of slowly building up human and physical infrastructure through the capture of the Chinese countryside's long-standing surpluses (Arrighi 1994, 2009; Hung 2015). As 'compensation' for the 'squeeze of rural surplus,' the Chinese 'party-state invested in agricultural infrastructure, basic education, and health care' in the rural cooperatives (Hung 2015, 47). While this strategy did not result in a spectacular growth rate, it did establish the conditions for Chinese capitalist 'opening up' to the world economy in the late 1970s and the subsequent take-off in growth. These conditions included the production of a relatively healthy and literate rural labour supply and the construction of extensive physical infrastructure, especially in rural areas. In combination with a set of other historical and geopolitical conditions (including the establishment of a US-led Cold War military economy in the region and the mobilization of an entrepreneurial Chinese regional diaspora), the Maoist period thus laid the foundation for the Chinese capitalist take-off (Hung 2015; Li 2016; Arrighi 2009).

An initial period of 'entrepreneurial capitalism,' centred on the agrarian initiatives of the Town and Village Enterprises, inaugurated the capitalist reforms of the Chinese economy in the 1980s. A period of state-led development followed that stressed the importance of the export sector. Most investment by the Chinese state since the 1990s has been concentrated along the urbanized coast, where most of the export-oriented industrial units are located. It is this export-oriented sector that has been the primary motor of Chinese economic growth since the 1990s. In 1985, the value of Chinese exports to the US totalled US$2.3 billion and to the rest of the world was US$27.3 billion. By 2013, the value of Chinese exports to the US was US$369 billion and to the rest of the world was US$2210 billion. This represents an 80-fold increase in export value to the world in about a generation—and a 160-fold increase in export value to the world's largest consumer market, the US (Hung 2015, 76). The booming export sector has raised millions of people out of poverty more rapidly than any other national economy has ever managed. However, this apparent success has also created problems of inequality within China as well as concerns about the burden of dependence on exports. The trajectory of Chinese capitalist growth since the 1990s, unlike with other export-oriented East Asian developmental states, has increased class, urban/rural, and interregional inequality within China, and the oversized influence of the export sector is also at the centre of a host of unsustainable balances in China and in the world economy (Hung 2015; Li 2016).

Because its growth strategy is predicated on exporting manufactured goods, the Chinese economy must continuously reinvest a substantial amount of its surpluses into manufacturing capacity. Political economic factors also exert pressure on the Chinese economy to maintain its current focus on export-oriented growth. The coastal industrial elites who exert considerable influence in the party-state and the managers of the influential and massive (but not relatively profitable) State-Owned Enterprises also favour the current mix of enormous outlay on fixed infrastructures and state support of the export sector. Minqi Li argues that the Chinese 'ratio of accumulation' (the ratio of net investment to profit) was around 40 per cent between 1990 and 2005, 50 per cent between 2006 and 2008, and has surged to 69 per cent since 2009. That is much higher than the US economy since 1995, which has come to rely heavily on financial rather than industrial accumulation (Li 2016, 86). Using a different metric,

Hung shows that fixed capital formation (net investment as a share of gross domestic product, or GDP) in China has skyrocketed from about 25 per cent in 1990 to above 45 per cent in 2011—even as it has dropped in other East Asian economies from about 32 per cent in 1990 to less than 25 per cent in 2007 (Hung 2015, 78). China's export dependence and massive investments in fixed capital work in tandem with the repression of domestic consumption, especially rural consumption. In 1990, private consumption as a share of GDP was just under 50 per cent in China. In 2011, this same figure hovered around 35 per cent. However, private consumption as a share of GDP in the other East Asian economies has actually risen steadily—from under 55 per cent in 1990 to just above 55 per cent in 2005 (78). All this is to say that the structure of Chinese over-accumulation is related to, but distinct from, the general East Asian model of state developmentalism.

This combination of factors has created a skewed capitalist growth machine perennially scrambling to figure out how to invest the massive surpluses accruing from the export sector. The classical Marxian assertion—that the structural tendency of capitalist accumulation is to create more surplus than opportunities to invest that surplus—is, therefore, apposite in the analysis of the historical geography of capitalism in China. Indeed, one sure sign that China is suffering from overinvestment is the fact that many of the loans for investment, distributed to Chinese firms through cheap credit, ultimately cannot pay for themselves—at least partly because there are simply not as many profitable opportunities to invest such massive surpluses. The theory of the spatial fix argues that responses to the contradictions of over-accumulation will take the form of spatial strategies such as shifting geographies of capital investment and sinking capital in long-gestation projects like physical infrastructures. The spatial fix offers a useful analytic lens from which to understand the infrastructure and investment politics of the Chinese state over the past couple of decades.

This structure of Chinese over-accumulation provides the necessary context to understand the spatial shifts in Chinese accumulation since the early 2000s. In essence, the Chinese state has attempted to guide and control the inevitable spatial fixes that come to the surface as a response to the contradictory structure of over-accumulation. The Chinese state has attempted to direct the spatial fix through two major policies—the 'go west' and the 'go out' policies. The former directs capital to operate in the relatively underdeveloped western frontier provinces of China, and the latter directs capital beyond China's shores to capital-hungry sites. Although 'go west' targets Chinese territory and 'go out' is directed across borders, both policies are a multiscalar state attempt at a controlled spatial fix to the contradictions of over-accumulation in the coastal boom towns of China.

At the intersection of these two policies is a planned massive energy grid that would draw on the rich natural resources of Xinjiang and, eventually, wean China off energy imports from Russia (Board of Investment 2015c). The envisioned integration of China's restive western provinces into a continental infrastructure project represents a geopolitical logic of pacification by the state. In this geopolitical imaginary, the 'backward' and ethnically diverse western regions of China—especially Tibet and Xinjiang—would be incorporated into modern development by the Han-dominated state (Yeh 2013). A fuller analysis than I am able to undertake in this chapter would more thoroughly and critically examine the conflict-ridden and antagonistic social space claimed by the Chinese state, particularly in relation to its western frontier regions and ethnic and religious minorities. For present purposes, however, it suffices to say that the 'go out' policies and the 'go west' policy—inaugurated in the early 2000s as a way to direct capital, infrastructure, and migrants towards the western interior—are beginning to dissolve into some larger infrastructural projects that integrate with Asia and beyond, but with China at the centre.

David Harvey (2001, 2003, 2006) derived the theory of the spatial fix from a structural understanding of the contradictory tendencies in the capitalist accumulation process. However, a concrete analysis of a specific historical-geographical conjuncture requires a more complex and nuanced understanding of social space. Infrastructural spatial fixes do not occur in smooth social spaces—rather, they must contend with the existence of contradictory and complex social spaces in particular places. In situations where social space is fragmented to such a degree that there is active resistance to state projects of incorporation, the state may resolve to secure the spatial fix through military means. Unfortunately, such action can bring about the increased fracturing of social space as well as deepen the core-periphery differences that led to resistance to state efforts in the first place (Akhter 2015b). A brief analysis of the CPEC project illustrates the urgency of attending to the historical-geographical specificity of social space and the manner in which this complexity shapes the geopolitics of the spatial fix.

China-Pakistan: 'Sweeter than Honey, Stronger than Steel'

As mentioned in the introduction, both Chinese and Pakistani state officials are given to hyperbole when describing the potential and significance of their renewed infrastructural relationship. Given the scope and scale of three CPEC projects, their optimism is not without foundation. Chinese investment is now pouring into Pakistan for numerous energy and infrastructure projects. On 22 November 2015, the Pakistan Board of Investment announced the signing of over 50 Memorandums of Understanding between Pakistan and China. These included agreements for the provisioning of Chinese governmental concessional loans for a motorway between Multan to Sukkur, a motorway between Havelian and Thakot, the East Bay Expressway for Gwadar port, and the Gwadar International Airport (Board of Investment 2015a).

The spectacular scale of CPEC should not distract from the fact that Chinese infrastructural capital and expertise has had a significant presence in Pakistan for some time now. Perhaps most prominently, China is widely thought to have supported the Pakistani nuclear programme (Small 2015). The Karakoram Highway, which already connects northern Pakistan to the Chinese city of Kashgar in Xinjiang province, was inaugurated in 1978, after more than a decade of coordination between Pakistani and Chinese engineers (Haines 2013; Kreutzmann 1991). There is also the pervasive influence of Chinese capital and expertise in the hydropower sector. As of 2013, 21 dams in various stages of completion were financed, built, or developed by Chinese companies. Two dams—the Diamer-Bhasha Dam and the Dasu Dam—are, respectively, the seventh and eighth largest dams by capacity currently being constructed in the world (World Energy Council 2015, 18) and, according to the website of the advocacy NGO, International Rivers, Chinese finance and expertise is involved in both. In Pakistan, the major Chinese financiers of dams include the China Development Bank, the China Exim Bank, and the Industrial and Commercial Bank of China while the more prominent builders and developers include the China Three Gorges South Asia Investment Company, the Dongfang Electric Corporation, and Sinohydro.

These flows of capital and expertise constitute a networked production of space in the contemporary infrastructural moment in the history of the Indus. The massive infusions of Chinese energy and infrastructural capital are intended to provide the conditions of capitalist accumulation within Pakistan and to integrate Pakistan into a larger Asian infrastructural space.

The port of Gwadar, near Pakistan's shared coastline with Iran, is envisioned as a Chinese window to the Arabian Sea. A rail/road network that stretches up through Pakistan and into China's western province will connect Gwadar to the city of Kashgar, Xinjiang.

While CPEC may thus be viewed as representing the creation of a smooth logistical economic space through state-of-the art transport and communication infrastructure, such a state space is never entirely smooth, frictionless, and bereft of any historical traces. On the contrary, social space is produced as heterogeneous, fractured, and contradictory over long periods of time and always in relation to the uneven regional distribution of political power over space. To come somewhat to grips with the complexity of social space in the areas of Pakistan where the most controversial CPEC projects will be located requires a brief synopsis of the historical geography of regional state formation.

The Fractures and Antagonisms of Social Space

This section provides a brief overview of the historical production of Pakistani national space as a militarized and antagonistic fracture between regions of the country. Secessionist and regional autonomy movements have had a significant presence in all of the underdeveloped provinces of the country for over half a century. In order to understand the militarized politics of uneven development in Pakistan, it is necessary to recognize the historically dominant role of Punjab—today, the most politically and demographically powerful province—in the process of state formation in Pakistan.

Under the British Raj, Punjab provided a steady stream of food and recruits for the army. This led to the cultivation of a rural Punjabi landed aristocracy aligned with military and civil elites: a triad that would powerfully shape the future of Pakistan (Jalal 1990). Indeed, the colonial 'Punjab school' of administration adopted in the region was a 'semi-military, despotic form of government' (Talbot 1988, 34), in which Henry and John Lawrence were two British brothers who essentially ran the entire region with the support of civilian and military officials who were invested with administrative and judicial powers. Even after the passage of the Punjab Code of Civil Procedure (1862) and the Punjab Laws Act (1872), authoritarianism was a hallmark of life in Punjab that stretched well into the twentieth century (Mathur 1966). In the 1937 elections, for example, only one quarter of the electorate included 'non-agrarian' castes (Talbot 1988).

Even predating the formal establishment of Pakistan, a fundamental contradiction existed between the liberal theories of legitimate democratic representation that Pakistan inherited from the British and population geography and the location of power. The 'Pakistan Movement' was spearheaded in British India by the Muslim League in explicit opposition to the all-India nationalism expressed by the Congress Party. And, arguably up until the 1940s, the Muslim League's most numerous and solid support was in parts of British India where Muslims were a minority. Headed by the lawyer Mohammad Ali Jinnah, the Muslim League was composed largely of what Hamza Alavi (1988) called the Muslim 'salariat,' that is, educated, middle-class, civil servants, who derived social standing from the scarce positions they occupied as 'native' subjects in the British Indian state. The Muslim salariat feared that the Muslim community would be overrun by the numerically superior 'Hindu community' in an independent India, and endorsed the slogan for 'Pakistan,' a nation composed of Muslim India, in the Lahore Resolution of 1940 (Jalal 1990).

This demand for Pakistan was not paradoxical as long as the Muslim League's position was read as a demand for greater representation of the Muslim population within a unified Indian constitutional set-up. It remains unclear exactly what Muslim League leaders had in mind, but, by 1946, 'Pakistan' had become a battle cry for creation of a separate state composed of the territories of British India with a Muslim-majority population. This strategy exposed a fundamental contradiction of political geography as it became clear that support for Pakistan came predominantly from Muslim-minority provinces that were decidedly not destined to become part of a new Muslim state. As Jinnah and the Muslim League soon realized, in order to be taken seriously as a representative of Indian Muslims, the powerful, populous, and prosperous Muslim-majority province of Punjab must be counted amongst their supporters.

Over the course of the 1950s, the elected branches of government in Pakistan lost power to the military and bureaucracy. It is tempting to dismiss the parliamentary activity that occurred between 1947 and the 1958 coup as a mere 'facade' that masked the reality of actual rule by the 'military-bureaucratic oligarchy' (Alavi 1972), but it is important to also recognize the difference between outright military rule and periods in Pakistani history where politicians and political parties have been operational. During the 1950s, a series of autocratic dismissals of provincial and federal cabinets—by bureaucrats, a strong and tightly centralized bureaucracy, and the active role of soldiers in multiple spheres of governance—were clear indicators of the relative weakness of political parties and politicians (Sayeed 1967).

The unelected bureaucracy and military are institutions of the state with deep roots in the province of the Punjab. In Pakistani political discourse, these powerful non-elected branches of the state, and especially the army, are referred to as 'the establishment.' On 7 October 1958, President Iskandar Mirza, a career bureaucrat, dissolved the newly minted 1956 Constitution of Pakistan, dismissed the national and provincial assemblies, and declared martial law. Mirza appointed General Ayub Khan as Chief Martial Law Administrator in the belief that the elite bureaucracy and the army would share power while continuing to exclude elected politicians. Three weeks later, General Ayub Khan cemented the place of the military as the dominant partner in 'the establishment' by dismissing Mirza from the presidency and sending him into exile. Ayub assumed the title of President for himself and implemented policies that further centralized power into a unitary state apparatus that exacerbated tensions between Pakistan's unevenly developed regions (Akhter 2015a, b). The most spectacular manifestation of this tension was of course the war of independence fought by Bangladesh (formerly East Pakistan), during which the occupying Pakistani army massacred and ravaged many thousands of Bengalis—including targeted attacks on women and intellectuals (Bass 2013).

This contradiction in the political geography of federalism continues to haunt Pakistan and takes the form of fear of 'Punjabization' of the country (Talbot 2002). Intersecting with this domination by Punjab is the role played by the armed forces in controlling the country's economy, foreign policy, and official nationalist ideology (Akhtar 2010; Siddiqa 2007). To sum up, the social space of the region that is current-day Pakistan is deeply heterogeneous and fragmented. Relations between its Punjabi-dominated central state and other regions are often fraught and militarized. Instead of a national-popular hegemony, then, what we see in Pakistan is a deeply antagonistic and resentful relationship between core regions of the country and their dominated peripheries. Any infrastructural intervention into this complex and historically formed social space will require a militarization of the space in order to secure the spatial fix there.

In other words, although both the Pakistani and Chinese states wax poetic about the space-smoothing effects that the CPEC projects will have in Pakistan, the deep-rooted social-historical structures in place generate tendencies towards the militarization of corridors and enclaves of logistical/

infrastructural space. Although it is very early in the implementation of these projects, one area where militarization is apparent is in the recruitment of military labour. True to its practice in other areas of the global South, Chinese infrastructure and logistics firms involved with CPEC projects plan to bring much of their technical and manual labour with them (see French 2014, for reportage on the African case). Many of these Chinese labourers will be working in economically depressed and politically restive areas of Pakistan, especially in the western provinces of Balochistan and Khyber-Pakhtunkhwa. As mentioned above, intellectual and political regional elites of these regions have often argued that state-led development projects in Pakistan favour the already-dominant province of Punjab and the already-rich sectors of the economy that are controlled by the military (including construction, real estate, and logistics). As a result, imported Chinese labourers and engineers have not infrequently come under attack by separatist militants in these parts (Akhter 2015c).

To ensure the safety of Chinese workers and companies working in these regions, the Pakistani army has taken the extraordinary step of raising a new division of troops dedicated to securing the CPEC. Dubbed the Special Security Division (SSD), this special unit will consist of nine thousand army soldiers and an additional five thousand paramilitary troops. Speaking at a press release on CPEC in 2016, members of Pakistan's defence establishment proclaimed a multidimensional approach to securing the Chinese spatial fix: 'Apart from security on land, the government has also taken relevant initiatives through the maritime security agency to protect the coast as well as through the Pakistan Air Force' (cited in Khan 2016). According to Khan, the initial reported cost of raising this security force is US$17 million. The federal government has also tasked provincial governments with raising supplementary security forces to assist the SSD. In April 2017, an official of the government of Khyber-Pakhtunkhwa announced that the 'government will purchase equipment, weapons, ammunitions and vehicles' for the securitization of CPEC, and that 'together with the salaries of the force's personnel, all these purchases are likely to cost the provincial government Rs 1.2 billion [US$1.1 million] annually' (cited in Dawn 2017). Instead of providing jobs for Pakistan's many millions of unskilled manual workers, or its many thousands of skilled technical workers, then, it seems that the greatest employment-generation effects of the CPEC may well be in the decidedly unproductive area of military labour.

The Chinese state's attempt to direct a spatial fix into the Pakistani landscape must thus confront the deep legacy of that region as a 'frontline state' for nineteenth- and twentieth-century global conflicts and imperial policing. Therefore, just as the historical and geographical particularities of Chinese state formation gave flesh to the abstract logic of capitalist over-accumulation, so the processes of colonial and militarized state formation in Pakistan contextualize the spatial fix in an actual historical geography. The structural analysis offered above suggests that CPEC may contribute to an exacerbation of the militarization and fragmentation of Pakistani national space. Ironically, this is exactly the opposite outcome to what is being presented about CPEC in the state and mainstream journalistic propaganda and which packages it as a 'game changer.'

CONCLUSION

Rather than a smooth space of transnational flows on a continental scale with the CPEC in Pakistan, a structural analysis suggests it is contributing to the reassertion of a deeply heterogeneous and fragmented social space that has evolved over several generations in the most underdeveloped regions of Pakistan. By deploying military labour to secure the spatial fix in Pakistan's peripheries, one effect is likely to be the deepening resentment of people living in these regions towards the Punjabi-dominated and militarized state. In other words, despite the rhetoric and ideology of infrastructure and logistics as connective and integrative, what this analysis suggests instead is the importance of remaining attentive to capital and social space as complex geopolitical forces or terrains of struggle in their own right. Rather than analyzing CPEC in terms of simple China-Pakistan interaction, then, this chapter has emphasized the importance of a structural and geographic-theoretical approach to understanding the geopolitics of inter-Asian infrastructural and logistical globalization.

This approach has several advantages over mainstream geopolitical analyses that attempt to understand the BRI as either a case of inter-state cooperation or inter-state domination. First, it allows us to situate state actors as embedded in a larger system of capitalist accumulation on a world scale. By understanding states, and indeed the state system, as both constrained and enabled by the imperatives and contradictions of capitalist accumulation, we help to demystify the notion of a state as an intentional agent interacting only with other similarly intentioned state actors. The necessity for a spatial fix to the contradictions of Chinese over-accumulation would exist regardless of the personalities of state elites that happen to be in the headlines now—and this has important implications for scholarly geopolitical analyses. Second, by examining social space as produced out of the historic interaction of a range of historical actors, we are better able to develop a multiscalar analysis of state power and social formations.

The borders of states do not contain smooth and undifferentiated spaces under the complete control of the state. On the contrary, the failures and successes of past state interventions (and the social reactions these interventions have catalyzed) shape the nature of social space as contested, fractured, and in some cases—antagonistic. Paying attention to the features of social space means maintaining a sceptical approach to the rhetoric and ideologies of state officials concerning the integrating and space-homogenizing effects of large-scale physical infrastructures and the logistical operations of capital.

By way of conclusion, it is important to note that the complexity and nature of any given social space is always contingent on particular place-based historical geographies. While this chapter outlines the contradictions and heterogeneity of political space in the peripheries of Pakistan, social space would be marked by different intensities, histories, and meanings in other parts of the world. Moving forward, a critical research agenda of the BRI, and of the processes of inter-Asian globalization more broadly, would benefit from historically and geographically grounded but rigorously theorized comparative analyses of the securitization of the spatial fix in other regional, scalar, and sectoral contexts. This chapter attempts to think through a useful theoretical framework to examine the geopolitics of the BRI from the perspective of the history and contradictions of capital accumulation, social space, and state formation. A more thorough analysis would necessarily engage in deeper research in the field and in the archives. Only by venturing in the direction of theorized and comparative analysis of the operations of capital can critical scholarship hope to develop explanations and political diagnoses that are adequate to the task of navigating the potentials and pitfalls of contemporary capitalism.

References

Agnew, John. 1994. 'The Territorial Trap: The Geographical Assumptions of International Relations Theory.' *Review of International Political Economy* 1, no. 1: 53–80.

Akhtar, Aasim Sajjad. 2010. 'Pakistan: Crisis of a Frontline State.' *Journal of Contemporary Asia* 40, no. 1: 105–22.

Akhter, Majed. 2015a. 'The Hydropolitical Cold War: The Indus Waters Treaty and State Formation in Pakistan.' *Political Geography* 46: 65–75.
———. 2015b. 'Infrastructure Nation: State Space, Hegemony, and Hydraulic Regionalism in Pakistan.' *Antipode* 47, no. 4: 849–70.
———. 2015c. 'Infrastructures of Colonialism and Resistance.' Accessed 29 April 2017. http://www.tanqeed.org/2015/08/infrastructures-of-colonialism-and-resistance/
Alavi, Hamza. 1988. 'Pakistan and Islam: Ethnicity and Ideology.' In *State and Ideology in the Middle East and Pakistan*, 64–111. Houndmills: Macmillan Education.
———. 1972. 'The State in Post-Colonial Societies Pakistan and Bangladesh.' *New Left Review* 74: 59.
Arrighi, Giovanni. 1994. *The Long Twentieth Century: Money, Power, and the Origins of Our Times*. London: Verso.
———. 2009. *Adam Smith in Beijing: Lineages of the Twenty-first Century*. London: Verso
Bass, Gary J. 2013. *The Blood Telegram: Nixon, Kissinger and a Forgotten Genocide*. New York: Knopf.
Board of Investment. 2015a. 'List of Pakistan-China MOUs.' Accessed 29 April 2017. http://boi.gov.pk/ViewNews.aspx?NID=%20762
———. 2015b. 'Pakistan, China can Produce Competitive Products for Global markets: Ahsan.' Accessed 29 April 2017. http://boi.gov.pk/ViewNews.aspx?NID=%20805
———. 2015c. 'China Pushes its Energy Network to Pakistan.' Accessed 29 April 2017. http://boi.gov.pk/ViewNews.aspx?NID=%20788. Accessed 29 April 2017.
Brautigam, Deborah. 2009. *The Dragon's Gift: The Real Story of China in Africa*. Oxford: Oxford University Press.
Chari, Sharad. 2015. 'African Extraction, Indian Ocean Critique.' *South Atlantic Quarterly* 114, no. 1: 83–100.
Cowie, Jefferson. 2001. *Capital Moves: RCA's Seventy-year Quest for Cheap Labor*. New York: The New Press.
Dawn 2017. 'KP to Form Force to Protect CPEC in its Area.' Accessed 29 April 2017. https://www.dawn.com/news/1325210/kp-to-form-force-to-protect-cpec-projects-in-its-area
Desai, Radhika. 2013. *Geopolitical Economy: After US Hegemony, Globalization, and Empire*. London: Pluto Press.
Desai, Vandana and Alex Loftus. 2013. 'Speculating on Slums: Infrastructural fixes in Informal Housing in the Global South.' *Antipode* 45, no. 4: 789–808.
Ekers, Michael and Scott Prudham. 2015. 'Towards the Socio-Ecological Fix.' *Environment and Planning A* 47, no. 12: 2438–45.
Flint, Colin and Peter J. Taylor. 2007. *Political Geography: World-Economy, Nation-State, and Locality*. Harlow and New York: Pearson Education.

French, Howard W. 2014. *China's Second Continent: How a Million Migrants are Building a New Empire in Africa*. New York: Knopf.
Chin, Helen, Fong Lau, Winnie He, and Timothy Cheung. 2015. *The Silk Road Economic Belt and the 21st Century Maritime Silk Road*. Kowloon: Fung Business Intelligence Center. https://www.fbicgroup.com/sites/default/files/The%20Silk%20Road%20Economic%20Belt%20and%2021st%-20Century%20Maritime%20Silk%20Road%20MAY%2015.pdf
Gallagher, Kevin P. 2016. *The China Triangle: Latin America's China Boom and the Fate of the Washington Consensus*. Oxford: Oxford University Press.
Glassman, Jim. 2011. 'The Geo-Political Economy of Global Production Networks.' *Geography Compass* 5, no. 4: 154–64.
Haines, Chad. 2013. *Nation, Territory, and Globalization in Pakistan: Traversing the Margins*. New York: Routledge.
Harvey, David. 2017. 'A Commentary on *A Theory of Imperialism*.' In *A Theory of Imperialism*, edited by Utsa Patnaik and Prabhat Patnaik, 154–72. New York: Columbia University Press.
———. 2006. *The Limits to Capital*. London: Verso.
———. 2003. *The New Imperialism*. Oxford: Oxford University Press.
———. 2001. 'Globalization and the Spatial Fix.' *Geographische Revue* 2, no. 3: 23–31.
Herod, Andrew. 1997. 'From a Geography of Labor to a Labor Geography: Labor's Spatial Fix and the Geography of Capitalism.' *Antipode* 29, no. 1: 1–31.
Hung, Ho-fung. 2015. *The China Boom: Why China Will Not Rule the World*. New York: Columbia University Press.
Jalal, Ayesha. 1990. *The State of Martial Rule: The Origins of Pakistan's Political Economy of Defence*. Cambridge: Cambridge University Press.
Jessop, Bob, Neil Brenner, and Martin Jones. 2008. 'Theorizing Sociospatial Relations.' *Environment and Planning D* 26, no. 3: 389–401.
Khan, Raza. 2016. '15,000 troops of Special Security Division to protect CPEC projects, Chinese nationals.' Accessed 29 April 2017. https://www.dawn.com/news/1277182
Kreutzmann, Hermann. 1991. 'The Karakoram Highway: The Impact of Road Construction on Mountain Societies.' *Modern Asian Studies* 25, no. 4: 711–36.
Lee, Chin Kwan. 2014. 'The Spectre of Global China.' *New Left Review* 89: 29–65.
Li, Minqi. 2016. *China and the 21st Century Crisis*. London: Pluto Press.
Liu, Weidong, and Michael Dunford. 2016. 'Inclusive Globalization: Unpacking China's Belt and Road Initiative.' *Area Development and Policy* 1, no. 3: 323–340.
Mathur, Y. B. 1966. 'Judicial Administration in the Punjab, 1849–75.' *Journal of Indian History* 44: 707–36.
Mezzadra, Sandro, and Brett Neilson. 2015. 'Operations of Capital.' *South Atlantic Quarterly* 114, no. 1: 1–9.

———. *Border as Method, or, the Multiplication of Labor*. 2013. Durham: Duke University Press.
Miller, Tom. 2017. *China's Asian Dream: Empire Building Along the New Silk Road*. London: Zed Books.
National Development and Reform Commission, the Ministry of Foreign Affairs, and the Ministry of Commerce. 2015. 'Vision and Actions on Jointly Building Silk Road Economic Belt and 21st-Century Maritime Silk Road.' Accessed 29 April 2017. http://en.ndrc.gov.cn/newsrelease/201503/t20150330_669367.html
Neilson, Brett. 2012. 'Five Theses on Understanding Logistics as Power.' *Distinktion: Scandinavian Journal of Social Theory* 13, no. 3: 322–339.
Neocleous, Mark. 2014. *War Power, Police Power*. Edinburgh: Edinburgh University Press,
Patnaik, Utsa and Prabhat Patnaik. 2016. *A Theory of Imperialism*. New York: Columbia University Press.
Samaddar, Ranabir. 2015. 'Zones, Corridors, and Postcolonial Capitalism.' *Postcolonial Studies* 18, no. 2: 208–21.
Sayeed, Khalid B. 1967. *The Political System of Pakistan*. Boston: Houghton Mifflin.
Schoenberger, Erica. 2004. 'The Spatial Fix Revisited.' *Antipode* 36, no. 3: 427–33.
Scott, James C. 1998. *Seeing Like a State: How Certain Schemes to Improve the Human Condition Have Failed*. New Haven: Yale University Press.
Siddiqa, Ayesha. 2007. *Military Inc: Inside Pakistan's Military Economy*. London: Pluto Press.
Small, Andrew. 2015. *The China-Pakistan Axis: Asia's New Geopolitics*. Oxford: Oxford University Press.
Smith, Neil. 2010. *Uneven Development: Nature, Capital, and the Production of Space*. Athens: University of Georgia Press.
Stevens, Andrew. 2015. 'Pakistan lands $46 billion investment from China.' Accessed 29 April 2017. http://money.cnn.com/2015/04/20/news/economy/pakistan-china-aid-infrastucture/
Talbot, Ian. 1988. *Punjab and the Raj 1849–1947*. New Delhi: Manohar.
———. 2002. 'The Punjabization of Pakistan: Myth or Reality?' In *Pakistan: Nationalism Without a Nation*, edited by Christophe Jafferlot. London: Zed Books: 51–62.
World Energy Council. 2015. *Charting the Upsurge in Hydropower Development*. London; World Energy Council.
Yeh, Emily T. 2013. *Taming Tibet: Landscape Transformation and the Gift of Chinese Development*. Ithaca: Cornell University Press.
Xinhua. 2015. 'Silk Road Fund's 1st investment makes China's words into practice.' Accessed 29 April 2017. http://english.gov.cn/news/top_news/2015/04/21/content_281475093213830.htm

CHAPTER 12

Becoming Immaterial Labour: The Case of Macau's Internet Users

Zhongxuan Lin and Shih-Diing Liu

As the oldest and last European colony in Asia, Macau seems to be an ideal target to illustrate the dynamics of 'logistical worlds' and 'labor of making a world region': Macau used to be a major regional trading hub connecting Asia and the world and a transit port for the infamous trade of coolies (indentured labourers) that shipped locals from South China to South American ports (Fung 1999). After losing its competitive edge to Hong Kong as a regional trading centre, Macau began to develop its gambling industry, with enhanced marine transport between Hong Kong and Macau to bring millions of gamblers from Hong Kong every year (Chan 2000). Thanks to spectacular economic growth since its handover to China in 1999, Macau's unemployment rate has dropped significantly against an excessive shortage of labour, resulting in a massive import of labour from mainland China and South-East Asia.

Z. Lin (✉)
School of Communication and Design, Sun Yat-sen University,
Guangzhou, P. R., China

S.-D. Liu
Department of Communication, Faculty of Social Sciences, University of Macau,
Taipa, Macau, P. R., China

© The Author(s) 2018
B. Neilson et al. (eds.), *Logistical Asia*,
https://doi.org/10.1007/978-981-10-8333-4_12

This historical background can be understood in terms of logistics through the control of labour and mobility, the changing relations between processes of production and circulation, and the remaking of urban and wider global spaces (Neilson and Rossiter 2014). The logistical perspective enables us to understand how labour and subjectivity and economy and society are tied to logistical worlds in particular contexts (Rossiter 2016). In an informational age, however, logistical infrastructure is no longer just about the ports, corridors, and territoriality; rather, it has become intensely connected to the massive volume of information produced by the Internet, media, and software that serve as emerging forms of infrastructure. To some extent, logistical worlds have become increasingly reliant on the immateriality of digital communication systems serving as the new pillars of global capitalism (Rossiter 2016).

Macau's peculiar position as a gambling capital not only demonstrates its potential as a nodal point of logistical worlds but also offers a local illustration of the transformation of labour in the Internet age. As the Macau government and the gambling sector have made huge efforts to promote informational, cultural, and creative industries, this chapter shifts the focus of analysis to the so-called immaterial labour characteristic of the Internet age. In particular, we employ the concept of immaterial labour to explore how Macau's Internet users become immaterial labourers and engage in new forms of communication, production, and collaboration.

Revisiting the Field

Immaterial labour is less about the production of things or 'finished products,' and more about immaterial services, intangible emotions, and communicative actions (Hardt and Negri 2004; Lazzarato 1996, 2004). This concept has risen to prominence with the expansion of the intellectual enterprise of 'autonomist Marxism' that delves into the transformation of capitalist production through the conceptual notions of 'empire' (Hardt and Negri 2000), 'multitude' (Hardt and Negri 2004), 'post-Fordism' (Virno 2007), and 'cognitive capitalism' (Vercellone 2007). Scholars like Lazzarato (1996), Hardt and Negri (2000, 2004), Virno (2004) and Terranova (2004) have investigated how immaterial labour, as an arguably more advanced form of productive power in late capitalism, reconfigures the network of post-Fordist production, distribution, and consumption processes.

More specifically, these studies have attempted to employ this theoretical concept to understand the post-industrial mode of capitalist production which has restructured labour practices and social formations through the enhanced intellectual, cognitive, communicative, and cooperative capacities of human labour (Hardt and Negri 2004; Lazzarato 1996, 2004; Virno 2004). As Hardt and Negri argue, 'the qualities and characteristics of immaterial production are tending to transform the other forms of labor and indeed society as a whole' (2004: 65). Following the Marxist line of thought, these studies have sought to use this concept to capture the new mechanism of capitalist exploitation and social control under the post-industrial labour regime (Lazzarato 1996).

In their conceptualization of the capitalist transition, Hardt and Negri (2000) argue that there has been a paradigm shift of production towards intellectual, immaterial, and communicative labour, and they identify the three most important categories of immaterial labour as follows: industrial production transformed by and incorporated in information and communication technologies; analytical tasks and symbolic work; and production and manipulation of affect. In this sense, immaterial labour also reconfigures the production of subjectivity (Virno 2004; Carah 2013). In the transition to this new production paradigm, everything becomes more productive—including labour subjectivities (Negri 1999)—because immaterial labour primarily involves the production, circulation, modulation, and manipulation of affect (Jin and Feenberg 2015), which is in itself intensely subjective (Dyer-Witheford 2001).

This autonomist approach, however, has been challenged by other scholars. For example, Caffentzis (2007) criticizes its excessive emphasis on the notion of 'immaterial' which, it is argued, offers up new jargon in the academic field while turning a deaf ear to social realities. Meanwhile, Dyer-Witheford (2005, 2009) further argues that although immaterial labour has become a prosperous research field, the concept itself remains unclear, vague, and reductive, running the risk of conflating too many different forms of labour while failing to adequately account for the nature and workings of the new labour mode. Besides the opaqueness of the concept, the romanticization of the subversive, emancipative, and rebellious potential of immaterial labour is also called into question by some scholars (Carls 2007; Dowling 2007; Nunes 2007). Nonetheless, and more importantly, most existing studies on immaterial labour remain limited to purely theoretical discussion and are obsessed with the definition, clarification, and distinction of the theoretical concept itself while ignoring the applicability of the concept in real circumstances to improving our understanding of the concrete practices of immaterial labour in context (Camfield 2007; McDowell 2009; Nolan and Slater 2010).

Following this critique, this chapter focuses on the concrete practice, operation, and performance of immaterial labour in the Internet age—with reference to the specific context of Macau's cyberspace—and inquires into three concrete questions concerning Macau's Internet users as immaterial labourers: what do these online immaterial labourers produce, how do they produce, and why do they produce?

In terms of the context of the Internet, scholars like Fuchs (2010, 2014), Arvidsson and Colleoni (2012), Andrejevic (2015), Robinson (2015), Comor (2015), and Zajc (2015) have analyzed immaterial labour in terms of 'user-generated-content' and 'audience labour.' On the one hand they consider Internet users as immaterial labourers embedded in complex processes of value creation and, on the other, they examine their exploitation within the informationalized, digitized, and networked capitalist economy. For example, Fuchs (2010) investigates how Internet users' time, attention, and content are sold to advertisers without any form of compensation and argues that this is a new form of labour exploitation that transforms Internet users into free labour in the service of informational capitalism. The effect, Fuchs claims, is that 'the exploitation of the commons has become a central process of capital accumulation' (190). Arvidsson and Colleoni (2012) also examine the way in which social media users' 'affective investments' are monetized by monopoly companies in the financial market by means of financial rents and stock market valuation rather than direct commodity exchange or the audience labour of Internet users. As a result, they argue, the value created and accumulated by Internet immaterial labour is ultimately possessed by the capitalists.

This specific focus on Internet immaterial labour is of great importance to our studies on immaterial labour in Macau. In the Macau case study, we examine the practices of immaterial labour with regard to the enclave's emerging informational economy at the turn of the century. Furthermore, we suggest that Macau illustrates the dynamics of the '[labour] of making a world region' in history and also demonstrates how the massive volume of information produced by Macau's Internet users has now taken the shape of new forms of logistical infrastructure in Macau—now a nodal point for the logistical worlds of the Internet age.

Macau's Internet Landscape

As the site of both an old Portuguese colony and a new Chinese Special Administration Region (SAR), Macau has integrated gradually into the circuit of global capitalism, and since the opening of the gambling market to foreign investment, it has evolved into an information society. When Macau officially launched its Internet service in 1995, its Internet penetration ratio of only 3 per cent was much lower than that in developed regions like Europe and the US. But in 2016, this ratio has reached to 88.6 per cent (DSEC 2017)—much higher than in either Europe (79.1 per cent) or the US (65 per cent), and more than twice the ratio of the Asia-Pacific region (41.9 per cent). Specifically, more than 77 per cent of Macau's Internet users have access to the Internet by mobile phone at any location, and almost 100 per cent of Macau's youth, aged from 18 to 30 years, are Internet users (DSEC 2017).

While this data suggests that Macau has become a developed information society, a gap exists between the rise of this information society and the development of its digital economy. As the only place in China where gambling is legally permitted, Macau's gambling industry has been further propelled by China's pro-market policy since the handover in 1999. In late 2006, Macau surpassed Las Vegas in terms of casino revenue, which accounted for about 80 per cent of its entire fiscal yearly income and upheld its status as the most powerful sector in the metropolis (Liu 2013). Due to this overreliance on casino capitalist production and gambling, however, other sectors—such as the information and digital industry— have not developed soundly in comparison. Although the Macau and the central governments have both sought to diversify Macau's economy—by supporting and promoting new economic sectors—the development of the Internet industry and, in particular, of professional Internet content and service providers, remains sluggish and lags behind. As a consequence, despite having a developed information society, Macau lacks a developed information industry that could engage its professional labourers, and its Internet economy must rely on some of its active 'amateur' labourers to produce local content on the Internet. This army of ordinary Internet users has become its immaterial labourers.

WHAT THEY PRODUCE

The first question concerns what exactly do immaterial labourers produce in Macau's cyberspace. Autonomist scholars have argued that the labour products of the new paradigm include service, knowledge, information, communication, relationship, and affect (Hardt and Negri 2004; Lazzarato 1996, 2004; Virno 2004). In our view, this claim is too general, so we want to find out what it means and how it manifests in a concrete context.

According to a local survey, one of the main reasons (86.5 per cent) and main activities (61.8 per cent) that Macau's Internet users go online is to obtain information (Cheong 2012). Although Macau has more than 20 newspapers and 10 TV channels—most of which have launched online versions including news websites, Facebook pages, and mobile applications—our ethnographic interviews found that these traditional media, and even their online versions, are usually not the first choice for most of Macau's younger respondents. For them, the traditional media outlets are 'too old and not cool' compared to a diversity of online contents, and thereby 'not attractive' to them. Moreover, they believe that the traditional media owned by the government and big firms tend to be biased and do not represent the public interest. Therefore, many local Internet users seek to rely on themselves to disseminate information by creating alternative media outlets online.

One of the most popular and representative dissident media sites identified during the fieldwork is the 'Macau Concealers' Facebook page, which was founded in November 2005. It has been critical of the Macau government and the pro-government newspaper *Macau Daily* and, with only one non-professional editor and no professional journalists, it mainly relies on its 87,593 followers' voluntary contributions for content. As the editor Alex explained:

> Without Internet users, we would not have Macau Concealers. We are not a professional news agency, and we don't even have professional journalists and editors. What is worse, we have no money. But we still want to give our voice to our government, and we think that is meaningful and necessary. So we have to count on Internet users completely. I thank them for their selfless contributions.

Another alternative media 'IMT Channel' Facebook page was founded on December 2010 by Tommy, who is also its only administrator. To manage the content, Tommy is usually on stand-by 24-hours a day, but even so he still finds it impossible to finish everything, and so he invites his 35,791 followers to voluntarily become 'journalists'—to gather, report, comment, and share news and information. As Tommy explained in the interview, 'I spent about two to three hours every day to administer IMT Channel, gathering news reports, editing and posting the news. Besides this routine work, I have to stand by in case of breaking news and responding to readers' urgent comments anytime. I have two smartphones to keep me online so that I can do my "job" anytime anywhere.'

Compared with traditional media, alternative media created and operated by Internet users tends to be 'deprofessionalized,' 'deinstitutionalized,' and 'decapitalized' (Atton 2002, 2009) and relies instead on ordinary Internet users' active participation and contribution (Baase 2008; Gillmor 2006). Some scholars claim that this signals a new paradigm of news production and consumption in the Internet age (Deuze 2011), in which Internet users, as non-professional volunteers, contribute their immaterial labour to produce and operate the alternative media. They are also consumers of the information they produce and, in this sense, have become enthusiastic 'prosumers' of alternative media (Bruns 2009; Rosen 2008), especially in a political context where official information sources and mainstream media are deemed untrustworthy. As the statement on the Facebook homepage of IMT Channel demonstrates, the aim of such alternative media may be to 'protect Macau citizens' public interest by reporting the news that the traditional media would cover up.' In this sense, the alternative media produced by Internet users challenge 'perceptions of the roles and functions of journalism as a whole' (Deuze 2003: 216), while also proposing challenges to the hegemony of traditional media (Bowman and Willis 2003). At the same time, it also offers an opportunity for ordinary Internet users to challenge and criticize the government (Fortunati et al. 2014). As Alex from Macau Concealers further explains:

> [W]e thank Facebook for providing such a platform for us to create our own media and give our voices. You know, the traditional media are mainly owned and controlled by the government; we need our own media, even though we need to contribute our own free labour. We believe in our worth, and we can walk on it. What we produce is not just the media, but another choice and opportunity for Macau Internet users.

In addition to news information, the other main 'product' Macau Internet users produce is a multiplicity of virtual communities based on Facebook. According to a survey by the Macau Association for Internet Research (MAIR), Facebook is the most popular and important platform in Macau (Cheong et al. 2011). Our fieldwork also identified many other virtual communities, such as 'Macau Aomen,' 'I am Macanese,' 'Macau People, Macau Affect, and Macau Affair,' 'Rainbow of Macau,' 'Macau LGBT Rights Concern Group,' 'Love Macau@520Macau,' 'true.lovemacau,' and 'Macao PPL,' all of which are established and operated by ordinary Internet users. For example, Macau Aomen is co-operated by its

32,587 followers, who constantly post videos, audio clips, photos, and songs about Macau. During the fieldwork, we rarely found any instances of rational discussion, deliberation, or professional information. Rather, the majority of information has to do with emotional feeling, pleasure, aesthetics, hobbies, and entertainment. Though usually simple in style and brief in content, these postings are widely liked, shared, and discussed by community members to express their sentimental feelings and memories of Macau. More specifically, the symbols and narratives produced by the members serve to construct a sense of 'virtual togetherness' (Bakardjieva 2003) and virtual communities (O'Connor and Mackeogh 2007). The space created by the collective engagement of Internet users is more like an affective space for nurturing a sense of nostalgia and belonging to an affective community. As Alice, the founder of the online communities 'Love Macau@520Macau' and 'true.lovemacau' stated, 'I serve as a volunteer for this Facebook page. I am really glad that these Facebook pages can attract so many Macau people to get together online to share our love for Macau. And I am proud that I can do something, although insignificant, for these communities.'

From the perspective of autonomist scholars, such as Hardt and Negri (2000, 2004), this kind of affective labour and community may embody a kind of exploitation. From the more traditional Marxist view, the form of affective communities may be expropriated by capital as 'the expropriation of the common' (Fuchs 2010, 188). But none has provided a satisfactory explanation for what we have observed in this digitized media space. According to our fieldwork, for most Macau's Internet users, exploitation is not an issue; rather, they have made use of social media to develop a strong sense of belonging, empowerment, and accomplishment and receive consideration and emotional support from other members of their affective community. In other words, this collective shared sentiment cannot be simply reduced to 'the expropriation of the common' because their members have found the culture created by their engagement more important than the issue of exploitation. As Alice further explains:

> [W]hat we create are not just one or two Facebook pages, but different online communities for different Macau netizens. To some extent, we are not just creating the community itself, but the relationship of Macau people. Yeah, definitely, we create the relationship based on the production of so many online communities. In this sense, we are not just producing online communities, but an online Macau.

In addition to the cases of alternative media and affective communities mentioned above, some of the online immaterial labour does turn into offline actions, which suggests that such labour does not always produce 'immaterial' products; rather, it has the potential to provoke 'material' actions in the real world.

A specific case of this in the fieldwork is the environmental campaign—'Saving the Egret Wetland'—which was collectively initiated and organized by several Macau Facebook groups such as 'Care for Birds, Love Macau' and 'Green Life in Macau.' The Egret Wetland is a popular destination for bird watching and, in 2012, the Macau government abruptly decided to construct a transportation centre on the wetland. Some citizens considered the development would damage the wetland ecosystem, and Facebook groups launched a campaign to protect the wetland habitat and wildlife. To demonstrate and convey their opinions, they established Facebook pages as alternative media to publicize their actions and gather signatures to petition the government. The Facebook pages were also turned into an effective community where members could produce posts, videos, and photos to express their affective solidarity. Examples of such postings included 'Protect Macau! Please do not be cruel to our home!' 'Keep the collective memory and feeling of the Macau people. This kind of memory is the root of Macau culture,' and 'Save Macau, save our home.' By 13 June 2012, they had collected 2280 signatures and produced 3243 comments, which led to the government giving up its original plan.

It is important that this kind of 'material' action (in the real world), arising as a 'product' of Macau's Internet users' immaterial labour, should be recognized. Macau has long been characterized as a small, quiet town, and Macau locals have long been designated as plain, simple, traditional, low-profile, and preferring harmonious relationships (Lam 2011). But what the Internet users produce may bring about change in the material world and even cause a change in Macau's political culture. As Rosa, one of the founders of the Facebook groups told us:

> Macau people, especially the youth, used to be accused of being politically indifferent because they have no interest in serious politics. What we have done both online and offline will be valuable if this kind of situation can change. Even the slimmest opportunity is worth a try. We will try our best to do whatever we can do.

How They Produce

Another deficiency found in the autonomist approach is that it overemphasizes the question of 'what'—what immaterial labour is and what it produces—on the macro and abstract level, while paying much less attention to the question of 'how'—in particular, how ordinary people become immaterial labourers and how they produce. Lazzarato (1996) has argued that 'immaterial labor constitutes itself in forms that are immediately collective, and we might say that it exists only in the form of networks and flows' (136). Indeed, the cases we observed are largely based on the collective mode of production that is inseparable from the cooperative network of Internet immaterial labour configured on social media platforms. However, our fieldwork found that there are always some key 'soft leaders,' like Tommy and Alice, who play a crucial role in the production process. In this sense, there are always leaders, representatives, intermediaries, and power relations among the 'collective' immaterial labourers in an informal and implicit hierarchy. In other words, the collective mode of 'peer production' or 'commons-based peer production' is not really as decentralized and disorganized as some would suppose but is in fact de facto managed by some soft leaders.

While these soft leaders may guarantee the efficiency, timeliness, and continuity of immaterial production, some problems may potentially arise in the online communities with which they are affiliated. For example, in January 2016, Tommy formally registered the 'IMT Channel' as an NGO which has, since then, become a platform to promote his personal images (personal portrait photos), contact information, and WeChat account on the heading column, with his images on the homepage rather than the 'public interest' platform he used to present. Alice and her communities have suffered similar problems—with easy-to-find traces of Alice's personal interests, preferences, and biases on the platform. In fact, Tommy and Alice are both social activists, with their own political values and affiliations, and their alternative media outlets serve their personal agendas as well as those of the political organizations to which they belong. In this sense, all the immaterial labourers involved in the production of the alternative media are sort of 'kidnapped' by particular persons and organizations, which is far from what Hardt and Negri (2000) optimistically and romantically assume to be the potential of communism (294).

Moreover, although autonomist scholars stress that the production of immaterial labour is 'collective' based on 'peer production' and 'commons-based peer production,' they do not really specify what kind of 'collective,' 'peer,' and 'commons' is embodied in productive processes. Our fieldwork seeks to ensure that the form of 'collective,' 'peer,' and 'commons' does not remain at an abstract level, as an anonymous 'mass' or silent majority. Instead, our research aims to embody the very concrete social relationships and networks of organizational members, alumni, classmates, fellows, schools, association members, and so on. Within Macau's context of 'associational society' (*shetuan shehui*), we found that most of the 'collective' immaterial labour we observed has specific organizational relationships in the offline world.

Macau has long been characterized as an 'associational society' and, by 2012, there were approximately 6000 associations in Macau, that is, 100 associations for every 10,000 people, or a density of 10 per cent—the highest in the world (Lou 2013). In addition to the quantity and density of its associational structure, the diversity of categories of associations also illustrates the importance of associations in Macau. The activities and functions of these associations are extensive and can be roughly divided into 13 categories that concern almost every aspect of the local residents' daily lives. Thus, citizens have been motivated to join specific associations, and it is within this particular context that the 'collective,' 'peer,' and 'commons' of Internet immaterial labour emerges in Macau.

With the development of the Internet, the existing offline associations extended their cultural influence onto online networks through the creation and operation of Facebook pages, as well the nurturing of various other affective communities online. In addition, increasing numbers of 'youth associations'—such as the Macau Youth Dynamics—have been inclined to appropriate social media and online networks as their preferred method of forming and organizing their communities (Shirky 2008). In the context of this 'associational society,' however, some of Macao's virtual communities still maintain some connection with associations in the real world. For example, all Facebook groups—such as Macau Concealers, IMT Channel, Love Macau@520Macau—have close links with particular social and political organizations in the offline world or, at least, their founders are affiliated with them. At the same time, the members of those social and political organizations are also active participants in alternative media communities.

Based on what we have observed, we want to argue that social relationships should not be viewed as the product of immaterial labour 'first and foremost' as some autonomist scholars have claimed (Lazzarato 1996).

Rather, predisposed social relationships may offer existing affiliations, networks, and resources for immaterial production. In other words, the production of relations itself is based on, and emanates from, some preexisting social relationships and networks that further expand and even reconfigure in the online space. For example, as alumni of a local middle school, Ivy, Judy, Cindy, Sharon, and Maggie voluntarily established a Facebook group 'Shcces Secrets' to connect with other alumni. In this case, the relationship of alumni as a virtual community provides the basis of their immaterial production online. As Cindy explains, 'We believed that we share some inherent lifelong relationships and experiences because we are all part of the same family of Shcces, and we try our best to do something [online] to help the "family members" to keep contact and maintain relationships.' After the launch of 'Shcces Secrets' Facebook page, the hosts regularly spent several hours a day—posting pictures, news, and articles, and encouraging members to 'like' the page and contribute as well. One month after its founding, more than 1000 alumni were connected to the Facebook page, but less time is now required in managing it—about 10–30 minutes per day. There has been an explosion of contents as the community expands, and more participants contribute their immaterial and affective labour. In another case, Helen and Jack provide free Internet labour because of their membership of the Macau Cinema Association—their work includes the recruitment of new members. Amongst the student interviewees, Liliana and her four classmates told us they established a Facebook group to work collaboratively on class assignments.

All of these examples suggest that participants in online groups contribute immaterial labour based on their group roles and membership as well as to develop and strengthen their networks. All have also become close friends as a result of their long-term collective engagement. Our fieldwork also finds that almost all of the alternative media and virtual communities we have observed in Macau have established social relationships and networks, which are reproduced and extended through social media. Indeed, although online immaterial labour has the potential to produce new relationships, its configuration is less likely to be affected by a completely new, transcendental and autonomous force.

Apart from the 'collective' form of immaterial labour production, our fieldwork also identifies that the 'individual' form of production is a central component in Macau's cyberspace. For example, CyberCTM, one of our ethnographic fields, is an online forum operated under CTM, a monopolistic telecommunication provider in Macau. CyberCTM, however, relies mainly on individual and non-professional Internet users to produce content and create value. For example, during the fieldwork, we observed five active members of CyberCTM forums, who spend at least one hour on the forums almost every day: Ada usually reads, comments, and posts news about stars and celebrities in forums on gossip, Jack on digital stuff, Doris on music, Derrick on finance, and Gary on cars. Everyone spends time doing different things on different forums, individually and separately, but this kind of individual labour as a whole creates, operates, and sustains the whole of the CyberCTM website. No matter whether done collectively or individually, we found that most of Macau's Internet users seem to voluntarily get involved in the online labouring process, and this resonates with what Terranova (2004) conceptualizes as 'free labor.' As these examples of Macao's social media groups show, the sustainability of the Internet depends on the substantial amount of free labour contributed by members and other Internet users.

WHY THEY PRODUCE

When discussing immaterial labour, autonomist scholars have frequently highlighted its exploitative aspects but without explaining why immaterial labourers would continue to actively produce under such precarious conditions of exploitation. Moreover, this approach has failed to engage with the complex motivation of labourers as to why some of them are willing to be exploited. At stake is the role of emotional need and value, namely affect, in the processes of immaterial production, for as Negri (1999) argues, affect has become a 'force of self-valorization' for immaterial labour. Furthermore, 'labor becomes affect, or better, labor finds its value in affect, if affect is defined as the "power to act"' (79–80). In our fieldwork, we found the concept of 'affect' as a force of self-valorization useful but suggest that it too needs reinterpretation and recontextualization. For Negri (1999) and others (Lazzarato 1996; Hardt and Negri 2000, 2004; Hochschild 1983), affect is understood more as the measure of the value of immaterial labour, and such value is usually considered as attached to the object rather than the subject of that labour. In other words, affective labour tends to be seen as serving to increase the satisfaction and happiness of the 'clients' rather than the 'labourers' themselves. But our fieldwork shows the opposite. We found that Macau's online immaterial labourers usually produce and appropriate affective labour for themselves and, at most, direct it to serving community members rather than any external clients. Such affect has become the main source of motivation and value itself for Internet labourers and constitutes positive affects produced by their immaterial labour—such as feelings of satisfaction, happiness, accomplishment, encouragement, as well as being appreciated, respected, and cared about. All work to energize and motivate the labourers and, in doing, animates and sustains their labour practice.

For example, Doris is an active participant in an online music community, 'Corner Music.' She spends much of her spare time contributing free labour to the development of the community and freely admits:

> We do not care (about exploitation), because we love music and we like this community. This kind of affect motivate us to contribute what we can do for other community members who love music as well, and I really enjoy this kind of feeling.

Peter, a local football player, explained why he devotes a lot of time to his Facebook page in terms of a sense of satisfaction gained from witnessing touching moments of solidarity with his team:

> I am satisfied by the positive feedback from my followers. When I see a message with encouraging words for Macau Football, I feel quite happy. Especially when someone encourages the players not to give up, I feel very touched. Sometimes, just because someone comments on my post, I feel encouraged. You know, playing football as a professional career is not easy in such a small city. This kind of feeling is just beyond any literal description. Someone appreciates you when you have done something, and you really enjoy the feeling of being appreciated.

Two other interviewees expressed similar feelings of recognition, accomplishment, and achievement. As one put it, 'When others read what I have shared and posted, if they are interested or find it meaningful, I feel like I have done something great. Especially when the posts I left are appreciated by others, I feel extremely satisfied and have a strong sense of achievement.'

In this sense, we suggest that love and other positive feelings as forms of affect do not simply create the 'power to act,' as Negri (1999) argues, but are a 'reward' that is constantly being generated and experienced by the participants. Our fieldwork also finds that the expression and sharing of affects such as loneliness, bitterness, sweetness, comfort, and encouragement have also become a main source of motivation for many Internet users. For example, on the student-run Facebook community called 'UM Secrets,' the motivation for most participants to contribute their free labour is that they can freely express their emotions in this 'secret' community: his/her happiness about falling in love; his/her love for someone else; his/her complaint about the pressure of study; and his/her loneliness and sadness. Of the interviewees who use Instagram, some expressed similar experiences. As Lily told us, 'Participating in Instagram means posting my daily life online and sharing it with my friends, as well as sharing the bitterness and sweetness around me.' Olivia, who has been an active participant on Instagram, recounted: 'When I go through my Instagram group, it is like a journey through which I read a lot of different stories and it gives me different feelings of sadness, happiness, and humor.' Some members of a local alumni group page similarly told us, 'We cannot turn back to our past, but you will always miss it. That is why we came here to share our memories.' As these accounts suggest, affect—as manifested in different forms of 'collective sentiment' (Hetherington 1998)—is a driving force that creates the value of 'we' and 'being-togetherness' in the production of immaterial labour and, as a result, works to produce and sustain the momentum of affective communities.

Conclusion

This chapter began with a discussion of Macau's historical specificity and its potential to serve as a nodal point in logistical worlds and explored the immaterial nature of digital labour. Through focusing on the salient characteristics of the development of immaterial labour in Macau's cyberspace, we explore how Internet users become immaterial labourers and investigate *what*, *how*, and *why* they produce in the emerging information economy. We contribute to the field by proposing a need for empirical study in order to understand the manifestations of immaterial labour practices—in particular what they do, how they work, and how they turn individuals into free labourers. These aspects of immaterial labour practice have previously not been examined by the autonomist approach. Moreover, in an attempt to highlight and address the gaps in this latter theoretical tendency, we have adopted a bottom-up, microscopic perspective that explores these questions in context by locating and examining the subjective experiences and agency of Macau's online immaterial labourers themselves. We also call into question the tendency of the autonomist approach to overlook the concrete local experiences of online group users.

We propose that a localized perspective is urgently needed to understand the particularities of labour practice in specific contexts of capitalist formation. Furthermore, as a SAR of China, the Macau case is not only useful for understanding aspects of Internet immaterial labour in situ, it also provides a laboratory for further investigation into Internet immaterial labour in China. This research resonates in a wider context with the theme of the 'labour of making world regions.' While Macau's local experience of immaterial labour practices may be partial, the findings of this study provide a useful base for comparable studies in other contexts.

References

Andrejevic, Mark. 2015. 'Personal Data: Blind Spot of the "Affective Law of Value"?' *The Information Society: An International Journal* 31, no. 1: 5–12.
Arvidsson, Adam and Elanor Colleoni. 2012. 'Value in Informational Capitalism and on the Internet.' *The Information Society: An International Journal* 28, no. 3: 135–50.
Atton, Chris. 2002. *Alternative Media*. London: Sage.
———. 2009. 'Alternative and Citizen Journalism.' In *The Handbook of Journalism Studies*, edited by K. Wahl-Jorgensen and T. Hanitzsch, 265–78. New York: Routledge.
Baase, Sara. 2008. *A Gift of Fire: Social, Legal, And Ethical Issues in Computing and Internet*. Upper Saddle River, NJ: Pearson Education.
Bakardjieva, Maria. 2003. 'Virtual Togetherness: An Everyday Life Perspective.' *Media, Culture & Society* 25, no. 3: 291–313.
Bowman, Shayne and Chris Willis. 2003. *We Media: How Audiences are Shaping the Future of News and Information*. Accessed 2 June 2017. http://www.hypergene.net/wemedia/download/we_media.pdf
Bruns, Alex. 2009. *Blogs, Wikipedia, Second Life, and Beyond: From Production to Produsage*. New York: Peter Lang.
Caffentzis, George. 2007. 'Crystals and Analytic Engines: Historical and Conceptual Preliminaries to a New Theory of Machines.' *ephemera: theory & politics in organization* 7, no. 1: 4–25.
Camfield, David. 2007. 'The Multitude and the Kangaroo: A Critique of Hardt and Negri's Theory of Immaterial Labour.' *Historical Materialism* 15, no. 2: 21–52.
Carah, Nicholas. 2013. 'Watching Nightlife: Affective Labor, Social Media, and Surveillance.' *Television & New Media* 15, no. 3: 250–65.
Carls, Kristin. 2007. 'Affective Labour in Milanese Large Scale Retailing: Labour Control and Employees' Coping Strategies.' *ephemera: theory & politics in organization* 7, no. 1: 46–59.

Chan, Sau-san. 2000. *The Macau Economy*. Macau: Publications Centre, University of Macau.
Cheong, Angus. 2012. *Surveying Macao ICT indicators*. Macao: MAIR.
Cheong, Angus, Xue Chang, and G. Lam Yip, 2011. *Macao Total Public Opinion Index 2011*. Macao: Macao Association of Internet Research.
Comor, Edward. 2015. 'Revisiting Marx's Value Theory: A Critical Response to Analyses of Digital Prosumption.' *The Information Society: An International Journal* 31, no. 1: 13–19.
Deuze, Mark. 2003. 'The Web and Its Journalisms: Considering the Consequences of Different Types of Newsmedia Online'. *New Media & Society* 5, no. 2: 203–30.
———. 2011. *Managing Media Work*. London: Sage.
Dowling, Emma. 2007. 'Producing the Dining Experience: Measure, Subjectivity and the Affective Worker.' *ephemera: theory & politics in organization* 7, no. 1: 117–32.
DSEC. 2017. *Survey on Information Technology Usage in the Household Sector in 2016*. Macao: DSEC.
Dyer-Witheford, Nick. 2001. 'Empire, Immaterial Labor, the New Combinations, and the Global Worker.' *Rethinking Marxism* 13, no. 3–4: 70–80.
———. 2005. 'Cyber-Negri: General Intellect and Immaterial Labor.' In *The Philosophy of Antonio Negri: Resistance in Practice*, edited by Timothy S. Murphy and Abdul-Karim Mustapha, 136–62. London: Pluto Press.
———. 2009. 'For a Compostional Analysis of the Multitude.' In *Subverting the Present, Imagining the Future: Class, Struggle, Commons*, edited by Werner Bonefeld, 247–66. Brooklyn: Autonomedia.
Fortunati, Leopoldina, Sakari Taipale, and Manuela Farinosi. 2014. 'Print and Online Newspapers as Material Artefacts.' *Journalism* 16, no. 6: 830–46.
Fuchs, Christian. 2010. 'Labor in Informational Capitalism and on the Internet.' *The Information Society* 26, no. 3: 179–96.
———. 2014. *Digital Labor and Karl Marx*. New York: Routledge.
Fung, Bong Yin. 1999. *Macau: A General Introduction*. Hong Kong: Joint Publishing (HK) Co. Ltd.
Gillmor, Dan. 2006. *We the Media: Grassroots Journalism by The People, for The People*. Sebastopol, CA: O'Reilly.
Hardt, Michael and Antonio Negri. 2000. *Empire*. Cambridge, MA: Harvard University Press.
———. 2004. *Multitude: War and Democracy in the Age of Empire*. New York: Penguin Press.
Hetherington, Kevin. 1998. *Expressions of Identity: Space, Performance, Politics*. London: Sage.
Hochschild, Arlie Russell. 1983. *The Managed Heart: Commercialization of Human Feeling*. Berkeley, CA: University of California Press.

Jin, Dal Yong and Andrew Feenberg. 2015. 'Commodity and Community in Social Networking: Marx and the Monetization of User-Generated Content.' *The Information Society: An International Journal* 31, no. 1: 52–60.
Lam, Iok Fong. 2011. 'Media, Identity and Civil Society: The Case of Macau.' Paper presented at the 8th International Seminar of 'Media and Environment.' Taipei: Fu Jen Catholic University.
Lazzarato, Maurizio. 1996. 'Immaterial Labor.' In *Radical Thought in Italy: A Potential Politics*, edited by Paolo Virno and Michael Hardt, 133–47. Minneapolis and London: University of Minnesota Press.
———. 2004. 'From Capital-Labour to Capital-Life.' *ephemera: theory & politics in organization* 4, no. 3: 187–208.
Liu, Shih-Diing. 2013. 'The Cyberpolitics of the Governed.' *Inter-Asia Cultural Studies* 14, no. 2: 252–71.
Lou, Sing Wah. 2013. 'Multi-categories and Pan-functions: The Diversified Development and Question Analysis of Macau Associations.' In *Annual Report on Economy and Society of Macau*, edited by Z. L. Wu and Y. F. Hao, 52–71. Beijing: Social Science Academic Press.
McDowell, Linda. 2009. *Working Bodies: Interactive Service Employment and Workplace Identities*. Hoboken, NJ: Wiley-Blackwell.
Neilson, Brett and Ned Rossiter, eds. 2014. *Logistical Worlds: Infrastructure, Software, Labour*. No. 1. Accessed 2 June 2017. http://logisticalworlds.org/wp-content/uploads/2015/04/535_UWS_Logistical-Worlds-digest-2014-v10-WEB.pdf
Negri, Antonio. 1999. 'Value and Affect.' *Boundary 2* 26, no. 2: 77–88.
Nolan, Peter and Gary Slater. 2010. 'Visions of the Future, the Legacy of the Past: Demystifying the Weightless Economy.' *Labor History* 51, no. 1: 7–27.
Nunes, Rodrigo. 2007. '"Forward How? Forward Where?" I: (Post-)Operaismo Beyond the Immaterial Labour Thesis.' *ephemera: theory & politics in organization* 7, no. 1: 178–202.
O'Connor, Barbara and Carol MacKeogh. 2007. 'New Media Communities: Performing Identity in an Online Women's Magazine.' *Irish Journal of Sociology* 16, no. 2: 97–116.
Robinson, Bruce. 2015. 'With a Different Marx: Value and the Contradictions of Web 2.0 Capitalism.' *The Information Society: An International Journal* 31, no. 1: 44–51.
Rosen, Jay. 2008. *Press Think: A Most Useful Definition of Citizen Journalism*. Accessed 2 June 2017. http://archive.pressthink.org/2008/07/14/a_most_useful_d.html
Rossiter, Ned. 2016. *Software, Infrastructure, Labor: A Media Theory of Logistical Nightmares*. New York: Routledge.
Shirky, Clay. 2008. *Here Comes Everybody: The Power of Organizing Without Organizations*. New York: Penguin Press HC.

Terranova, Tiziana. 2004. *Network Culture: Politics for the Information Age*. London: Pluto Press.
Vercellone, Carlo. 2007. 'From Formal Subsumption to General Intellect: Elements for a Marxist Reading of the Thesis of Cognitive Capitalism.' *Historical Materialism* 15, no. 1: 13–36.
Virno, Paolo. 2004. *A Grammar of the Multitude: For an Analysis of Contemporary Forms of Life*. New York: Semiotext(e).
———. 2007. 'General Intellect.' *Historical Materialism* 15, no. 3: 3–8.
Zajc, Melita. 2015. 'The Social Media Dispositive and Monetization of User-Generated Content.' *The Information Society* 3, no. 1: 61–7.

CHAPTER 13

Follow the Software: Reflections on the Logistical Worlds Project

Brett Neilson

The chapters in this volume slice across a project known as 'Logistical Worlds: Infrastructure, Software, Labour.' Sited in the cities of Athens, Kolkata, and Valparaíso, Logistical Worlds began as an attempt to examine intersections of infrastructure, software, and labour in three shipping ports whose geopolitical fortunes were being transformed by China's logistical expansion. In practice, however, the project's focus shifted—in part due to issues of access to some of the facilities in question: particularly the container terminal at Piraeus port in Athens that is run by the Chinese state-owned enterprise COSCO, but also owing to a stall in port extension initiatives at Valparaíso port in Chile. As a result, the research widened to encompass the hinterlands of these shipping ports as well as the formation of zones, corridors, and commodity production and trade routes that link to the facilities under investigation.

This widened research focus is evident in the chapters in this volume that centre on the project's second research investigation in the eastern Indian city of Kolkata (formerly Calcutta). Although some of these chapters deal directly with the Kolkata port—with its diffuse set of docks, locks, and navigational

---

B. Neilson (✉)
Institute for Culture and Society, Western Sydney University,
Parramatta, NSW, Australia

© The Author(s) 2018
B. Neilson et al. (eds.), *Logistical Asia*,
https://doi.org/10.1007/978-981-10-8333-4_13

devices which span the Hooghly River between Haldia in the south and Kidderpore in the north—others branch out to its hinterland. These include a study of the 'floating' city of Siliguri, a key logistical choke point in Kolkata's hinterland, and a critical analysis of attempts to situate Kolkata port within wider logistical connections between China and India. Apart from Kolkata, other chapters reach west to Europe and east to Hong Kong and Macau to register the changing stakes in logistics and power—related to infrastructure, labour, trade, geo-economics, geopolitics, and space—that are redefining Asia and its regionalisms in the twenty-first century.

Logistics provides the unifying thread across these diverse sites. By observing and reflecting on the capacity of logistics to negotiate spatial, cultural, social, and economic heterogeneity, the chapters stage a complex run of encounters. They seek to understand the role, shape, and inner logic of logistics in 'making a world region' while also examining the diverse experiences, contexts, and outcomes for labour within them.

Rather than reiterate and round off the arguments and observations made by the individual authors, in this concluding chapter I seek to consider their analyses in the context of the wider research conducted for the Logistical Worlds project. Clearly, some of the dynamics and connections examined in this larger ambit are already implicit in the movement between the chapters that focus on Kolkata port, other infrastructural installations such as those at Piraeus, or those that investigate the politics of transnational logistical corridors.

A major challenge for Logistical Worlds was to remain sensitive to the complexity of relations within and between regional locations while conducting research at the chosen sites. Partly this was a problem of project design, which we (I use this pronoun to describe the core group of project researchers including Ned Rossiter and Ranabir Samaddar) met by staging 'research platforms' that afforded opportunities for collective investigations to be guided by local researchers at particular sites as well as for the movement of researchers between and across them. Combined with digital methods of research planning, organization, and dissemination, this approach provided a conceptual and technical collaboration model that coordinated researchers from different locations and disciplines around targeted empirical investigations (see Kanngieser et al. 2014). In doing so, however, we also faced the challenge, both conceptually and thematically, of tying the resulting empirical investigations together. Despite the broadly stated interests—in logistics, software, China's global expansion, shipping ports, and labour regimes—there was also the question, therefore, of how to organize and steer the research so that it could be assembled in a coherent way.

Meeting this challenge was complicated by the fact that, while the Logistical Worlds research focused on globally articulated processes and technologies, elements of the research were located at particular sites. In other words, although we were interested in the role and effects of logistical practices and technologies in specific locations, we were highly aware that these practices and technologies were not exclusive to them. Consequently, our investigations were not restricted simply to the study of location-specific phenomena but also sought to understand how processes and technologies enable, organize, and integrate disparate logistical practices globally. This meant that conducting a multisite study was fundamental as it enabled the researchers to move from site to site while also facilitating ongoing conversation between them by means of digital media.

More experimentally, we collaborated with Ilias Marmaras, Anna Lascari, and others in the production of a digital game (see cargonauts.net) to simulate and increase our understanding of the often 'black-boxed' software routines essential to the logistical control of global production and labour regimes. Such design efforts were not mere add-ons to the more traditional grounded practices of observation, interviewing, participation, or discourse analysis. Rather, they were an essential precondition for a situated engagement with logistical operations to proceed from an understanding of the infrastructural realities that underlie and enable them.

The logistical coordination of production and distribution on a world scale is often associated with what management guru Peter Drucker described in 1965 as 'the whole process of business' (quoted in Cowen 2014, 32). From the perspective of the contemporary empirical study of logistics, tracing this 'whole process' would be an unending research task spanning multiple supply chains and production networks. Pragmatically, the Logistical Worlds investigation had to be limited, and this led to the strategic decision to focus on three shipping ports, their hinterlands, and the corridors and trade routes that connect to them. Thinkers who have engaged critically with the project have commented that while 'the port has been privileged as the site of logistical operations,' much cargo handling 'has been displaced from ports to warehouses and distribution centres' (Gregson et al. 2017, 383). In reality, however, our study encompassed not only warehouses and distribution centres but also facilities such as railways, cable landings, and data centres.

In any case, the aim was not simply to trace the movement of cargo and, in this respect, our approach differed from that employed in many other multisited studies that have examined the intersections of politics and economy across diverse geographical scales. A time-honoured method for integrating such studies is to 'follow the commodity,' and this is indeed effective as a way of tracing material connections across diffuse global sites, regardless of whether they are conceived according to the model of the supply chain, commodity chain, production network, or business function. Another way of integrating such multisited research is to 'follow the money' or to privilege the investigation of financial flows, whether this involves the employment of a value chain model, attention to the reproduction of capital, or the tracing of specific actors and networks.

In Logistical Worlds, however, we sought to 'follow the software,' that is, to work from the perspective of the large technical systems that undergird the logistical coordination of production and distribution and to gain insight into the frictions and connections that enable 'the whole process of business.' Software, of course, is its own specific kind of commodity and new versions of it must constantly be generated in order that they can be sold, even if only as a service for a limited time period, to produce profits. This is an economy of updates that encloses firms, governments, and individuals within system dependency. Software is also a kind of fixed capital that enables different kinds of production processes, and which must be updated and developed by workers (in this case, coders but also an extensive number of early adopters, or beta users, that test software for bugs prior to general release) in order to maintain or shift its use value. Following the software has thus involved for us a complex combination of following commodities and following financial flows, which, in the context of our study, enabled an interrogation of specific operations of capital associated with logistics. Our decision to follow the software was not simply intended to expose the infrastructural conditions that shape logistical operations within and across sites, for while hardware may seem just as relevant and even more imposing in its material presence than software, the latter is certainly no less material or consequential for its being coded in programming languages and mediated through computing systems.

Our interest in following the software was rather a matter of locating and testing the limits of its operability. Logistical visions of production and transport are shaped by a desire for seamlessness that is, as almost all of the recent critical work on logistics has highlighted, shattered by the realities of labour, technology, and terrain. Software controls the movement of physical freight in logistical systems but, far from producing an integrated 'code space' (Kitchin and Dodge 2011), it consists of a patchwork of platforms, routines, and plug-ins whose interactions and mismatches tend to limit as well as enable that control. In the logistics industries, most software employed is proprietary and thus based on code that is unavailable for scrutiny or modification. That applies to all systems regardless of whether they are terminal operating systems, data information systems, haulage planning and operation systems, border security and customs systems, enterprise resource planning systems, or popular communications software such as Microsoft Messenger or Skype.

While customization routines adapt systems to local circumstances, and electronic data interchange standards promise to ease the passage of information across platforms, the reality of achieving interoperability is often elusive. Therefore, part of our approach in Logistical Worlds has been to deliberately search out situations where the operability between software systems breaks down, or where software systems encounter forms of labour or organization that intervene in their fields of operation, or even disrupt or displace the processes of coordination or control they are supposed to exercise. By addressing our investigations to such mismatches and limits—or what Ned Rossiter (2016) calls 'logistical nightmares'—we proceeded according to the principle (or perhaps rather the hunch) that these situations would guide us to sites where there was material evidence of labour struggles, social inequalities, or cultural translation. In the remainder of this chapter, I explore three such sites in order to critically analyze how following the software has shaped and articulated our empirical research across the cities of Athens, Kolkata, and Valparaíso.

SOFTWARE MISMATCHES IN PIRAEUS

Observing the port of Piraeus from its surrounding hills in 2014, it was possible to imagine that the discrepant arrangement of shipping containers on either side of the container terminal was an effect of the different terminal operating systems in operation. Piraeus is an ancient port that has survived until the present day, partly due to its strategic location on shipping routes between Asia and Europe. As Greece's premier port, it consists of a ferry port and car terminal as well as the container terminal, and the latter has evolved into an important transshipment hub where containers are unloaded from large vessels from China and transferred onto smaller feeder craft that carry them to other Mediterranean ports.

As Pavlos Hatzopoulos and Nelli Kambouri detail in their chapter in this volume, in 2009 more than half of the container terminal at Piraeus was leased by the Greek government to a local Greek subsidiary of COSCO Pacific Ltd, itself closely aligned in its leadership structure to the Chinese state-owned enterprise, COSCO Group. Since August 2016, COSCO Pacific (now known as COSCO Shipping) has become the majority shareholder in the Piraeus Port Authority (OLP) and has been granted management of all port services until 2052. During the Logistical Worlds research in 2014, the container terminal was divided into two parts: Pier 1, which was run by OLP, still under majority Greek government ownership; and Piers 2 and 3, both of which were run by Piraeus Container Terminals (PCT), the local COSCO subsidiary. Pier 3, which is equipped with post-Panamax cranes capable of unloading the largest vessels from China, was recently constructed by PCT. These changes to the ownership and management of the port were brokered during the harsh conditions of the economic crisis in Greece, under the oversight of the Hellenic Republic Asset Development Fund (HRADF). This agency was formed in 2011 to direct the privatization of Greek public assets as mandated under the terms of the first bailout package by the 'institutions' of the so-called troika—the European Commission, the European Central Bank, and the International Monetary Fund. The situation pertaining in 2014, when the container terminal was divided between OLP and PCT, contributed to the software mismatch explored in this section of the chapter.

There are many ways of describing the differences between the operations on the OLP and COSCO sides of the port at the time of our research. As Hatzopoulos and Kambouri explain, a dominant narrative about them is civilizational in logic—suggesting that a process of 'Chinification' was underway at PCT and pointing to changes in the labour regime that reflected a diminution of rights once accorded to Greek dock workers, and a concomitant downgrading of Greece in the hierarchy of nations. Logistical Worlds, however, questioned this 'Chinification thesis' on multiple fronts. For a start, we asked to what degree the supposed heroic masculinity of the Greek dock worker was sustained by relations of gender, race, and nation (Kambouri 2014). We also explored how the labour regime on the COSCO side of the port was enabled by patterns of subcontracting and precarization that were exacerbated by the economic crisis in Greece (Parsanoglou 2014). Furthermore, we compared the labour situation at PCT with that pertaining in Chinese ports, such as Yantian and Shekou, where strikes had recently occurred. This allowed us to suggest that what was at stake in the COSCO operation was not a process of Chinification but a transformation of labour relations and processes made possible by a unique collision of Chinese management practices and European social conditions. All of this, however, said very little about logistical practices and their infrastructural mediation and conditions, and it was with this question in mind that we examined the software systems in operation in Piraeus.

At the time of research, the OLP site was running the NAVIS SPARCS N4 terminal operating system—a product of the California-based company NAVIS and the most widely disseminated software of its kind in the world, particularly in North American and European ports. By contrast, PCT was running CATOS—a product of the South Korean firm, Total Soft Bank (TSB)—which is predominantly used in Asian ports, including on China's east coast, which are among the largest container ports in the world. Kap Hwan Kim and Hoon Lee (2015) detail the features of these different operating systems. They note that N4 is 'standard package software' that is 'distributed and patched to customers through a version control,' while CATOS 'has different features from a package software and additional development effort may be necessary for the application to a specific customer' (61–2). Among the 'advanced functions' of CATOS, Kim and Lee list the ATC Supervisor which 'controls job orders for unmanned yard equipment in real time and performs advanced automatic job-scheduling' (64–7). An article published in *World Cargo News* in 2010, at the point when the contract between TSB and PCT was first announced, lists this feature as among those engineered into CATOS for its deployment at Piraeus: 'TSB will install the CATOS system, including its web-enabled operation management, Electronic Data Interchange (EDI) and statistics modules, plus its ATC Supervisor system for the monitoring and control of unmanned equipment, comprising the ATC Supervisor Server (ATCS) and ATC Supervisor Client (ATCC)' (*World Cargo News* 2010, 31). Although a similar module can be patched onto N4, and CATOS is theoretically interoperable with its US-made counterpart, these specifications point to capabilities lacking in the OLP-operated part of the port in 2014. Knowledge of these software discrepancies provides intelligence regarding the technical conditions underlying the productivity gains made at Piraeus—which handled 3.74 million 20-foot equivalent units (TEU) in 2016 as opposed to 880,000 in 2010 (*China Daily* 2017)—as well as associated changes to labour processes and relations on the PCT side of the port.

PCT is not the only container terminal in Europe running CATOS. The system has also spread to ports such as Bilbao, Malaga, Valencia, and Gothenburg, which have an interest in more easily accommodating vessels arriving from China. COSCO's interest in Piraeus is often presented as a 'shining star' in China's Belt and Road Initiative (Spiliopoulou and Anagnostopoulou 2017), which extends China's economic and geopolitical interests through building of trade routes, infrastructural investments, financing of development, and cultural exchange. The widening adoption of a system like CATOS attests to unforeseen consequences of this initiative which, in this instance, generates political spaces and administrative routines, not through state-driven grand strategy but rather by means of what Keller Easterling (2014) calls 'extrastatecraft,' or the making of polity through infrastructural and technical systems that operate in parallel, partnership, or rivalry with the state (Neilson 2017).

On the ground, these software developments mean that the two sides of Piraeus port exist in different logistical 'worlds,' or at least this was the case at the time of our research in 2014. Such circumstances are not unusual in contemporary shipping ports which, as Daniel Olivier and Brian Slack (2006) explain, have been subject to a process of 'terminalization,' meaning that many ports host multiple terminals run by different private interests, and these terminals in turn form part of global corporate and infrastructural networks. With these arrangements, there can be a greater degree of integration between the operations of geographically dispersed terminals run by a single firm than there is between adjacent terminals located in a single port. The capacity of a software system like CATOS to operate simultaneously across many such geographically dispersed terminals intensifies this tendency.

In recent years, Total Soft Bank has been promoting the capacity of CATOS for 'process mining' operations that can increase port efficiency (*World Cargo News* 2013). Process mining is a form of data mining that uses event logs, which detail past equipment moves, to break down and recompose the container handling process in order to reduce equipment downtime. The technique takes into account the investment made in different pieces of infrastructure and configures operations so that quay cranes, for instance—which, are the most expensive and least flexible terminal hardware—are less idle than other equipment. Process mining functions more effectively in automated or semi-automated environments, such as those running TSB's ATC Supervisor, which generate more detailed event logs. In their contribution to this volume, Hatzopoulos and Kambouri describe how the logistical governance at work at the COSCO terminal in Piraeus uses these techniques to subordinate labour to the control of 'machinic idleness.' Detailing the network topology of CATOS, they show how a combination of rigid and flexible hierarchizations and process mining techniques makes human bodies, and labour, into appendages to computational processes.

Although process mining can measure and analyze human-machine interfaces, it generates data for human resource allocation but not for the remuneration of labour based on performance. This arrangement makes it necessary to have labour constantly on call in order that supply chain gluts or frictions can be smoothed over by the deployment of extra workers at peak times. In this respect, the labour regime at PCT differs from that on the OLP side of the port, where a shift system was in place, at least until August 2016, and wage arrangements were set by negotiation between the Greek government and the troika. PCT works with a more flexible wage system; labour is allotted on a day-shift basis, and the payment of overtime rests on management decisions.

Writing with Ursula Huws, Hatzopoulos and Kambouri argue that these technical conditions produce 'a complex mixture of flexibility and strict discipline.' The specification of these conditions, they suggest, makes the concept of precarity 'too general' for 'analysing existing labour relations and the power exercised over the labouring bodies of workers in Piraeus port' (Hatzopoulos et al. 2014, 22–3). Yet while software discrepancies between OLP and PCT provide an analytical angle from which to discern differences in the labour regimes on either side of the port, there is danger in assuming a too neat and totalizing vision of these technical systems. The view from above, mentioned at the beginning of this chapter section, shows containers on the COSCO side of the port tightly stacked while those on the OLP pier are more dispersed. Yet something else is learnt when approaching this division between the OLP and PCT from the ground, as we did in 2014 when we were guided through the OLP terminal by a union member.

As we came up to the line separating Piers 1 and 2, which is redolent in so many civilizational, economic, and territorial ways, we were informed that containers passing from one side of the port to another were not accompanied by a data flow from CATOS to N4 (or vice versa) but were rather logged by hand. Despite the capacity of both platforms to communicate with the other through electronic data interchange protocols, here was an example of a breakdown in interoperability between software systems. This slippage from the digital to the analogue marks a brief and temporary transition to a governance regime that no longer abides the logic, search, and calculation that defines the event logs of terminal operating systems. Prone to human error, the media of paper files also enables a kind of autonomy of labour that briefly and paradoxically asserts itself against the real-time media of information command and control. Moreover, in this case, the abstract mismatch between software systems is shadowed and physically reinforced by a line on the ground that acts at once as a site of division and connection between container terminals with very different commercial imperatives, labour regimes, administrative relations to states, patterns of ownership, and connections to international trade routes and logistical hinterlands. Movement between the software and territorial dimensions of the two piers provided a way of examining the deep inseparability of digital and material worlds in logistical practices, even as our investigations in Piraeus moved on to examine other relevant infrastructural installations, such as the railway connecting the port to a planned logistics hub at Thriassion and the wider network of transport corridors envisioned under the European Union's TEN-T programme (see Grappi in this volume). The tensions and discrepancies surrounding the line between the two sides of the port also animated our research as we moved further afield to Kolkata and Valparaíso, and institutional and commercial arrangements at Piraeus port shifted to make the COSCO subsidiary the majority shareholder of OLP by August 2016.

Off Grid in Kolkata

On a preparatory research visit to Kolkata in 2015, we conducted an experiment in a container park close to the Netaji Subhas docks (NSD), the main container receiving facility of the Kolkata Port Trust, adjacent to the Kidderpore docks on the Hooghly River. Randomly recording numbers on containers stacked in the yard, we proposed to track the global movement of these boxes for a period of six months by entering the numbers into a container tracking site that was freely available on the Internet—track-trace.com. The site aggregates container moves across more than 125 companies, and the results upon first checking were interesting. Although many of the containers whose numbers were entered into the site could not be tracked, many of those that were able to be tracked had their most recent location registered as Singapore. This told us two things. First, that the arrival of the containers in Kolkata had not been registered in a software system that was interoperable with the particular tracking application we were using. Second, that in the case of Kolkata we were dealing not with a major transshipment hub like Piraeus but with a smaller port that was supplied with containers transshipped through Singapore.

Following ten containers over the next six months, we noticed that five reappeared in Singapore, several weeks later, and then continued on diverse journeys around the world. Another travelled through the port of Colombo in Sri Lanka before moving on to China, and then on to the US. The remaining four containers remained off grid, presumably staying in the Kolkata yard or moving elsewhere in the subcontinent.

The experiment randomly selected only a few containers and thus cannot be considered to offer statistically significant results. Nonetheless, the fact that containers became invisible on the track-trace system once they entered the subcontinent tells us something important. It points not only to the uneven distribution of logistical technologies around the world but also to the extent to which the logistical vision of seamless communication and transport is a fiction tied to hopes for a future global capitalism that is less compromised and contradictory than its present manifestations. What we learnt from this experiment would be relevant for a wider investigation, in and around Kolkata: that the movement of goods and information into and out of digital regimes of calculation and knowability is crucial for the organization of labour in the logistical industries and the articulation of logistical modes of production to other operations of capital.

In this way, the research in Kolkata challenged us to explore the limits of software's control over logistical worlds. This concern with limits was paradoxical given the prominence of software in visions of shining India and the importance of Kolkata in schemes such as China's Belt and Road Initiative. It also explains why many chapters in this volume that focus on logistical operations in Kolkata do not engage centrally with the question of software. If, in Piraeus, the breakdown of interoperability between software systems was less a matter of technical possibility than of institutional and economic arrangements, in Kolkata we encountered situations where the reach of software into logistical systems was either partial or overlooked.

Such an observation, however, does not mean that the presence of sophisticated logistical and software systems in Kolkata port, or other infrastructural installations in the city and its hinterland, should be ignored. The question is rather about how, in the Indian case, the logistical production of space, time, and value proceeds at the disjunction between software economies, material infrastructure, and hard labour. During a visit to the Haldia Dock System—the bulk and container port operated by Kolkata Port Trust some 120 kilometres downstream from Kidderpore—we encountered a port manager whose handling of logistical information confirmed what we had learnt from our earlier experiment regarding the physical movement of containers beyond computerized inscription. One of the questions we posed to the manager, who was responsible for assigning teams composed of permanent and temporary workers to the loading and unloading of vessels, was about the terminal operating system installed at the port. He didn't know how to answer. Clearly the interface between container operations, software, and the bulk of the labour employed at the docks to move raw materials, such as coal and magnesium, was not critical enough to draw his attention. Although Haldia moves only about 30 per cent of the containers that pass through the Kolkata port system, it was nonetheless surprising to us that the manager seemed unaware of the software dimensions of the port's operations. Eventually, one of his offsiders volunteered that they use an out-of-the-box solution supplied by Tata Consultancy Services—MACH (Marine Container Handling System). Later in the interaction, the manager showed us a port manifest printed out from the computer records but accompanied by a page scribbled with his own handwriting, to which he referred in citing the vessels that would come and go.

Here, as in the instance of the paperwork used to log the passage of containers between the two sides of Piraeus, was a clear example of how port operations can go off grid. The official's scribbled notes reclaimed a predigital media of governance in which the inventory of goods and productivity of labour are not tethered to real-time systems of search and calculation. What we learn from these incidents is that software—far from being a promised scenario of total control—tends to be constrained by logistical practices as they unfold on the ground. As the research progressed in Kolkata, it became clear that the port's operations are shaped by dated, often colonial era forms of infrastructure and austerity economies that are complemented by media and storage infrastructures whose historical cultures and technical properties remain resistant to any straightforward translation into the spatial and temporal parameters of contemporary logistics.

Digital media and software systems, however, are not absent from the Kolkata port. In his contribution to this volume, Iman Mitra details Kolkata Port Trust's various terminal upgrading exercises as well as its investments in navigational technologies to assist the passage of ships up the Hooghly. Mitra explores correspondences between the institutional logics animating these investments and speculations regarding the port's involvement in space-making exercises, resource grabs, and real estate rentier economies. His findings concerning the role of financialization and public-private partnerships in assuring the port's viability confirm Laura Bear's (2015) claim that 'short-term improvised austerity technologies predominate on the river' (49). Bear argues that since the liberalization of the Indian economy in the early 1990s, Kolkata Port Trust, which remains a public institution, has functioned according to a 'logic of fiscal deficit' and 'low-tech solutions to generating revenue' (48). While there is no doubt that the port's facilities are decrepit and rusted, we should be cautious about reading inefficiency or labour exploitation off this façade, just as we should resist reading efficiency or favourable labour conditions off the shiny new equipment on Pier 3 in Piraeus. In the classical political economic sense, the presence of exploitation is not gleaned from experience or ethnographic observation. It requires, rather, a calculation based on the proportion of unpaid surplus labour a worker performs for their employer to the necessary labour required to produce the value equivalent of their wage. In this respect, Kolkata port certainly measures up to its more high-tech equivalents, although the prominence of contract work that exceeds the wage relation based on measured time is certainly a factor that needs to be accounted for on the banks of the Hooghly.

The Kolkata Dock System includes the Netaji Subhas docks (NSD), which have been upgraded recently to accommodate five container berths, and the adjacent facilities at Kidderpore that handle bulk cargo such as pulses, timber, fertilizer, and iron and steel. Further upriver, there are liquid petroleum facilities at Budge Budge. The container berths at NSD are equipped with cranes and yard equipment that transfer containers onto trucks that approach the docks through narrow laneways. A recent *Economic Times* report, which notes an increase in turnover, attributes this change to 'technological upgradation in container handling' (Himatsingka 2017). This expansion of container operations at NSD is largely the result of a 2014 privatization drive that granted a ten-year contract to Bharat Kolkata Container Terminals, a local subsidiary of Singapore's PSA International.

The private contractor, Bharat Kolkata Container Terminals, agreed to mechanize three container berths with mobile harbour cranes (adding to two container berths that were already using ships' cranes in their operations) and supply, maintain, and operate yard equipment such as rubber-tyred gantry cranes, reach stackers, and tractor trailers. A corollary to this process has been an outsourcing of labour through this same private contractor, adding to conditions of precarity already prevalent since the 1990s.

In addition, and despite the technological fix, NSD remains a ramshackle affair, that is, as Bear (2015) describes, 'a stretch of barren ground dominated by a queue of trucks waiting to load containers, old yellow dock labor buildings empty of people and a vast ruined clock tower' (88). But the recent upgrade has consolidated Kolkata's position as a 'feeder' port for Singapore, at least in so far as container operations are concerned, because the latest contracting arrangements have incorporated NSD into the global network of terminals run by PSA International, a corporate entity created after the privatization of the Port of Singapore Authority in 1997.

On the Kidderpore docks, there has been no comparable upgrade. Much of the unloading of goods from vessels proceeds manually and remains seemingly unchanged since 1944 when the photographs of the 'human conveyer belt' included in Kaustubh Mani Sengupta's chapter were taken. Migrant workers, both men and women, board the vessels to stitch commodities into sacks; then male workers carry the loaded sacks on their backs to storage or transfer sites. In Kidderpore today, workers who transfer the sacks from ships are still paid according to the weight they handle and this arrangement seemingly flouts the global dictates of the 'container revolution,' which was based on a logic of modularization that made volume rather than weight the object of measure in the shipping and stevedoring industries. Yet, while such techniques of labour measurement may not apply to the container yards at NSD, we should not conclude that labour relations are highly discrepant in these sites. As Bear documents, austerity labour regimes have dominated across the Kolkata port system since the 1990s, when fiscal approaches to debt—which accumulated largely due to the costs of dredging—displaced expectations that the port's finances could be negotiated through political relationships. In practice, this fiscal approach spurred a process of 'manpower reduction and casualization' (Bear 2015, 39). Older unionized dock workers with permanent public-sector positions were not replaced and an increasing portion of dock work is now performed by temporary workers who have to take on non-unionized jobs—on short-term, often only one-day, contracts managed by brokers. These arrangements extend across NSD, where the introduction of containerization in the 1990s drove the rise of casualized work. At the Kidderpore docks, according to Iman Mitra and Mithilesh Kumar (2016), the hiring of migrant workers maintains elements of the *sardari* system—a 'classical jobber system' by which a *sardar* (or head) 'recruits around fifty workers under him and puts them to work under a contractor' (28).

A further aspect of the intensification of the labour process at Kidderpore is related by Mitra and Kumar who note that whereas, previously, the sacks could hold 100 kilograms of commodities, they now have a capacity of only 50 kilograms. While this change reduces back strain and the risk of injury for workers, it also increases the time and pace at which they must work to unload a vessel and earn the same amount of money. Mitra and Kumar observe that such an 'innocuous and as mundane tool of work as the sack can perform the same task of intensifying and informalising work as advanced automation on the docks is capable of doing' (28–9). This story is worth remembering with regard to our earlier point about the reach of software into logistical systems at Kolkata port. If increased labour intensity is often associated with technological changes of the kind we have observed at Piraeus, and to a lesser extent at NSD, then it is important to recognize that such adjustments can also be achieved through 'low-tech' fixes, such as, in this case, the introduction of a smaller sack made of plastic threads.

Working across these examples, we see how the intensification of labour has been accompanied by its diversification—in terms of technical composition—and, also, heterogenization—with regard to its legal and social regimes of organization (Mezzadra and Neilson 2013, 87–93). What matters is not some zero-sum correlation between software control and labour intensification, but how differing degrees, rhythms, and technologies of intensification work off each other. The spectre of workers carrying sacks beneath the shadow of modern container equipment signals something more than the juxtaposition of production modes that Ernst Bloch (1977) glossed as the 'synchronicity of the nonsynchronous'—the spatial coexistence in the same time period of historically heterogeneous practices and social formations. What becomes evident is rather the extractive element of logistical operations—their ability to generate value by organizing the exploitation of social cooperation in ways not directly attributable to the fraction of capital that benefits from this extraction. In the case of the sack-carrying workers at Kidderpore, for instance, there is the imposition of the *sardari* system, whereby a contractor (sardar) recruits and administers the labour. It is like the financial logic of the derivative, which creates value by speculating on underlying assets that are never themselves transacted. While life and labour become the raw materials on which the logistical edifice is built, at the same time, over this same edifice, they seem to float or glide as assets that are expendable or exploitable precisely because they do not register on the screens or in the processors where value comes to be measured. In the instance of the Kolkata Port Trust, which operates under fiscal conditions of austerity, these extractive operations apply not only to the organization of labour, through various degrees of subcontracting and informalization, but also to the drawing of value from land assets leased out under diverse and often temporary terms (again, see Iman Mitra's contribution to this volume). Yet, despite the complexity of these arrangements, logistical systems remain operative, albeit in ways not found in the promotional literature of shipping lines, distribution centres, software firms, and supply chain management companies.

Logistics, in other words, exceeds the formal dimensions of production and distribution. Modes of capital accumulation driven by logistical operations enter the many and diverse tributaries of informal systems of economy and labour and draw on seemingly residual systems of governance and social relation. These systems are not simply forms of social or economic activity that are prior to or *not yet* subdued by capital. They are also produced *from within* capital, through its prospecting initiatives and through resistance to its logic. In the case of the Kolkata Port Trust, they are also produced from within state institutions which, under austerity conditions, display a sharp conflict between their redistributive and extractive functions. The transition from the formal, computational governance of supply chains to informal and makeshift practices off the grid also signals how the inner workings of logistical systems elude inspection. Whether human or machine, logistical knowledge evades those not immanent to the situation of operation, and following the software supplies a means of both gauging the reach of digital technologies into logistical practices and assessing the limits of this knowledge.

Where the Trace Ends

As mentioned above, the planned Logistical Worlds research in Valparaíso, Chile, was diverted due to the stalling of port extension initiatives. When the project was first proposed in early 2012, plans were afoot to expand Valparaíso's container receiving facilities so they would be capable of receiving the largest vessels arriving from China. The hope was to revive the fortunes of this nineteenth-century port, whose shipping traffic had declined with the opening of the Panama Canal in 1914.

At Valparaíso, there were three expansion plans. The first involved the extension by 120 metres of Berth 3 at Terminal 1 or Terminal Pacífico Sur (TPS)—a terminal that moves 90 per cent of Valparaíso's container traffic.

The second involved the enlargement of Terminal 2 or Terminal Cerros de Valparaíso (TCVAL)—a member of the Spanish Group OHL—to add a container terminal capable of serving simultaneously two super-post-Panamax craft (or vessels too large to pass through the Panama Canal even after its 2016 expansion). The third initiative involved the construction of a new Terminal 3 or Terminal Intermodal Yolanda—a large-scale container terminal to the north of the existing port facilities. Despite the presence of impressive architectural drawings and plans to link these facilities to logistical routes that would connect Valparaíso to wider Latin American markets—for instance, through a low-altitude tunnel running through the Andes—only the first of these initiatives has been realized. The second is permanently stalled and the third is the subject of political controversy.

The reasons for these delays in the expansion of Valparaíso port are complex. For a start, the berth expansion at TPS has given that facility the capacity to accommodate two post-Panamax craft, and a recent upgrading at the Port of San Antonio to the south has added container processing capacity with which Valparaíso may not be able to compete. Second, the logistical difficulties posed by moving containers over the Andes, and the unlikelihood that Valparaíso could evolve into a transshipment hub, mean that the port will probably not become a strategic port of call for super-post-Panamax ships arriving from China. Third, the expansion of the Panama Canal has meant that larger ships can now reach Latin America's east coast. And fourth, the status of Valparaíso as a UNESCO World Heritage Site has boosted citizen struggles over the port's development— particularly the expansion of Terminal 2, the building of Terminal 3, and the construction of a shopping mall between the sites. Thus, although China remains Valparaíso port's number one import partner, and number two export partner (*South China Morning Post* 2015), it is unlikely that container operations will increase massively to accommodate trade from China as they have, for instance, in Piraeus.

These developments, however, did not deter us from following the software at Valparaíso port. We investigated, for instance, the introduction of the Silogport port community system by the Valparaíso port authority (EPV). This electronic platform was designed by the Spanish firm INDRA, in collaboration with EPV, to connect the multiple software systems run by public and private organizations that make up the port community: shipping terminals, shipping firms, trucking companies, the Chilean railways, the *Zona de Extensión y Apoyo Logístico* (ZEAL) or dry dock operated in the hills above the city by EPV, the customs service, and even the Ministry of

Finance—which has made its SICEX system for the automatization of foreign trade processes interoperable with Silogport. We also tracked how the berth expansion at TPS, and subsequent installation of three new post-Panamax cranes, was accompanied by a switch of terminal operating software from an in-house system to NAVIS SPARCS N4. As a press release on the NAVIS website makes clear, this change had implications for labour because—by utilizing a vehicle routing optimization plug-in called Prime Route that coordinates with N4—TPS was able to 'reduce the average terminal tractors from 5.5 to 4.5 per gang' (NAVIS 2016).

In our visits to TCVAL, we found further evidence of how software shapes labour processes and relations. On a wall near the entrance to the terminal, computer printouts were posted of the names of workers rostered to certain shifts. We were told that this list of names corresponded with biometric information linked to a fingerprint scanning device on which the workers had to check in and out of the terminal. Far from merely providing a means to identify workers, this system turned out to have important resonances with the labour history of Chilean ports and contemporary labour struggles in the sector.

The computer-generated list of names both continues and subverts an earlier port labour system known as *la nombrada*. Active in Chilean ports until the time of the military dictatorship, this system involved workers being rostered to shifts by trade union leaders who would call their names in hiring halls. Supposedly, the system involved named workers passing on the right to work to another worker, called the half chicken, who could then pass it on to another, known as the quarter chicken, and so on, with wages being divided. This aspect of the system was purportedly dismantled in 1981, when the junta's *Plan Laboral* opened the ports to non-union labour and introduced day contracts. But in an increasingly neoliberalized environment, port operators found this old labour supply institution amenable because, especially when controlled by local trade union leaders open to negotiation, *la nombrada* offered a tool of labour flexibilization and atomization (Rojas 2016). Combined with a new safety registration system, the flexible use of this system led to a situation where, eventually, 80 per cent of workers in Chilean ports were *eventuales* or day labourers hired on contracts.

Beginning in 2010, however, the hiring halls through which the system was administered became sites of rank-and-file organization, which led to a series of rolling strikes in which dock workers used newfound associational capacities to apply leverage at key economic chokepoints. Unsurprisingly, port operators turned against *la nombrada* and were able to successfully dismantle it, for instance, in San Antonio, where a larger permanent workforce was hired. The electronic *nombrada* we observed at TCVAL—but also instituted at TPS—was essentially an attempt to maintain the flexibility of the system while reducing union control in the name of transparency and 'social peace' (TCVAL 2016). In this instance, software and database protocols offered a new institutional configuration in labour hiring processes, coordinating them with biometric technologies, and providing more stability and benefits for regular casual workers in return for keeping the overall system flexible.

As our research advanced at Valparaíso port, it quickly became apparent that an investigation of China-driven logistical expansion in this part of the world could not limit itself to the port, or indeed the city, of Valparaíso. There were two main reasons for this. First, the logistical expansion driven by containerized imports from China was not confined to Valparaíso but had to be understood in relation to grander, parallel developments in San Antonio port, 60 kilometres to the south. San Antonio—with its ambitious plan for port expansion and one container terminal that has four post-Panamax and two super-post-Panamax cranes already installed—has effectively outstripped Valparaíso's capacity (Labrut 2015). As both ports serve populations in the Santiago Metropolitan Region, their competitive relations contribute to the formation of a single logistical complex or triangle.

Valparaíso thus forms a node within a wider logistical arrangement and any study that engages with the infrastructures that support its maritime industries must necessarily spill across the city's limits. The second reason why our research needed to extend beyond Valparaíso concerns the port's role in the export of copper to China. Chile is the world's largest copper producer and China is its primary customer. Once exported to China, copper is stockpiled, refined, and used in a variety of industrial applications, including the manufacture of computing and telecommunications hardware that is essential to today's logistical industries. Representing just under 50 per cent of Chilean exports, copper is mainly mined in the country's north and exported through ports such as Antofagasta. In Valparaíso port, copper is the fourth largest export—behind fruit, wine, and other comestibles—and most of it is refined and shaped into ingots or cathodes, depending on the degree of purity the metal has obtained. In recent times, China has preferred to import bulk copper concentrate over the refined product (Home 2017), and Chile has matched this preference by developing new technologies for containerizing concentrate in sealed boxes that are then exported from the northern ports. However, the refined copper trade remains a significant part of Valparaíso's turnover, and this locates the port within a complex mining logistical chain that is a primary source of its connection to China and guarantees its importance within electronics production networks that enable the expansion of digital media.

To research Valparaíso's position in this China-centred copper export trade, we decided to extend Logistical Worlds to trace the production of copper cathodes, the highly refined form of the metal that is favoured for electronics applications and traded as a commodity on the metal exchanges in New York City (COMEX), London (London Metal Exchange), and Shanghai (Shanghai Futures Exchange). This involved visiting the point of extraction at the Andina open pit mine (4000 metres above sea level in the Andes), the control room that coordinates the mine's extraction and processing activities, and the Ventanas smelter (40 kilometres to Valparaíso's north), which refines the copper mined at Andina into cathodes using an electrolytic process of which gold and sulphuric acid are by-products. All facilities are run by CODELCO, Chile's state-owned copper mining company, and the Logistical Worlds team had excellent access to processing plants and computer rooms. We also visited the Center for Mathematical Modelling at the Universidad de Chile, which works with CODELCO on big data projects that seek to build efficiency into the copper supply chain by, for instance, modelling seismic activities in mines or designing wearable technologies for miners.

What became clear in this research was the role of software in distancing the commodity and the worker from the business of copper extraction. Although multiple software systems are involved, each is active in a kind of ritual that separates the metal from its earthly origins. The mine is abstractly and topographically divided into 'blocks' of earth in which the concentration of copper has been predetermined by prospecting activities that make use of electronic sensors attached to drill heads. Theoretically, it should be possible to trace the extracted material from such a 'block' through the concentration process, which takes place in a giant cavern in the side of a mountain at the Andina mine, and on to the smelting of the cathode—the final product—an extremely pure sheet of copper, approximately 1 centimetre thick and 1 square metre in area. Although such tracking and tracing systems are currently not in place, they are becoming increasingly possible to implement as automation is built into the production process, for instance, by the use of remote sensors, laying of fibre optic cable, real-time tracking of personnel, use of semi-autonomous vehicles, control and monitoring of equipment, and installation of 'ruggedized' computers.

In a sense, though, the making of such a trace, which would give the copper provenance, is precisely what the production process works against. As the metal is separated from the extracted ore, it undergoes a process of valorization by which it is abstracted into an anonymous and fungible commodity that becomes tradable on futures markets. At the same time, the refining process converts the raw material into a chemical element—with the atomic number of 29—that could have been mined anywhere in the world. At Ventanas, the management is proud of the extremely high-quality cathodes the smelter produces and does not lament the fact that their origins cannot be read off their atomic structure once they are loaded into containers and shipped to China. They also recognize, however, that the abstraction and valorization that animate this production process has a price.

That price is environmental degradation, for the area surrounding Ventanas has been declared *una zona de sacrificio* due to the contamination from heavy industry that has built up in the region since the 1960s. The smelter's emissions of sulphur dioxide and arsenic have controversially contributed to environmental devastation and the poisoning of Ventanas' communities. If copper is the symbol of Chilean modernity, the 'toxic lives' of the populations living in the vicinity of Ventanas—including the so-called *hombres verdes* or green men, ex-employees of CODELCO who have 'greenish lacerations produced by chemical reactions on their bodies' (Tironi and Rodriguez-Giralt 2017, 95)—reflect its shadowy and damaged underside.

Like the port sector, Chile's mining industry has recently been a hotbed of labour struggles led by temporary contract workers (Rojas 2016). CODELCO has been at the centre of these struggles with major strikes organized against it in 2007–09, and again in 2015 under the direction of the Copper Workers' Confederation (CDC), a grassroots union initially composed of outsourced and flexible workers neglected by the state and established trade unions. Like the dock workers who used the institutional form of *la nombrada* to press their case, the force of protest in the mining industry rests on the strategic position of these workers within an industry on which the Chilean economy is dependent. Increasing automation in mines like Andina is likely to pose a new front of struggle for these workers in the coming years—a struggle that will have to account not only for outsourcing of work to private labour contractors but also the organizational force of software in displacing the human from the machinery of capital. In this respect, following the software becomes not only a means of animating and coordinating research but also a political method relevant to the organization of struggles across industries and sectors, including but not restricted to those narrowly associated with logistics. These struggles are likely to evolve around multiple edges where logistical practices and software systems converge to facilitate the extractive operations of capital—regardless of whether the object of extraction is a raw material like copper or patterns of labour and social cooperation susceptible to data prospecting and valorization. The peculiar and powerful strand of neoliberalism that has developed in Chile allows us to see this nexus clearly.

Conclusion

The concerns that stream across the three sites of the Logistical Worlds project are multiple and cross-hatched. In this chapter, I have highlighted the role of software in logistical practices, concentrating on its relevance—for labour processes and relations, as well as capitalist processes of extraction, abstraction, valorization, and accumulation. Each site has attracted a different emphasis. In Piraeus, the mismatch between terminal operating systems on either side of the port reflects the different labour regimes in these facilities as well as the allegories of civilizational conflict that are often (falsely) called up to explain these differences. In Kolkata, the limited reach of software into logistical systems and their associated labour regimes shows why software can never be understood as a device of total control and how it operates in parallel and rivalry with other technologies of monitoring, calculation, and control. In Valparaíso, software provides an institutional frame that seeks to abstract economic processes from their historical and environmental contexts—whether these be labour regimes such as *la nombrada* or the various forms of dirt, danger, and toxicity associated with the mining and processing of copper.

Doubtless these are highly schematic summaries of what we have learned by following the software in (and beyond) each site, and there are possibilities to apply the knowledge and insights gained from research in any one site to the others. Indeed, the platform method of research we pursued in conducting the Logistical Worlds project encouraged—indeed, necessitated—such a transfer of knowledge and methods by moving researchers across locations. The results were often messy, but they also allowed us to observe common dynamics in the operations of capital across the sites, for instance, those associated with austerity, extraction, and the flexibilization of labour.

Earlier in this chapter, I mentioned that our interest in the uses and effects of logistical practices and technologies in specific locations was conditioned by an awareness that these practices and technologies were by no means restricted to the sites where research was conducted. Such a realization does not mean that our ambition to move and apply the knowledge gained, through research across sites, licenses a process of extrapolation or generalization that issues in the making of global conclusions about logistics. An important element of our findings concerns the ways in which local conditions determine to a substantive extent how logistics operates, including implications for labour regimes, political negotiations, and wider schemes of development, modernization, and international relations. At the same time, it is important to recognize that logistics works, across different contexts, with a persistent logic or rationale that seeks to build efficiency and interoperability into processes of transport, communication, and economic transaction.

Paradoxically, this rationale is not one of spatial or institutional homogenization. Rather, as Anna Tsing (2009) emphasizes, it thrives off the negotiation of diversity. In other words, logistics does not merely turn preexisting diversity to its own ends but also produces diversity, creating or revitalizing niche segregations, for instance, through the building of zones, corridors, and other specialized topographies that advise and stimulate economic performance. A purely logistical site is probably an exception, as debates about the Tennessee Valley Authority in the United States (Selznick 1949) or Soviet industrial towns (Allen 2003; Collier 2011) have shown. Nonetheless, it would be a mistake to emphasize how logistics adapts to and benefits from local circumstances at the expense of examining its capacity to make worlds.

Returning to the theme that animates this volume—the role of logistics in the making of world regions—we can observe the uneven reach of China's logistical expansion across the three sites in question. In Piraeus, there are indirect patterns of Chinese ownership in COSCO's gradual takeover of the port. In Kolkata, the forging of logistical connections to China is present in background policy frames such as India's Act East policy and China's Belt and Road Initiative. However, the predominant political and economic reality that structures logistical operations in the Kolkata port, as described by contributors to this volume, is postcolonial capitalism. This particular variety of capitalism is shaped by a subcontinental experience of empire, colonialism, and accumulation different to what unfolded historically in China. In Valparaíso, China looms as a crucial and distant trading partner whose appetite for copper—and increasingly fruit, salmon, and wine (Quiroga 2017)—fuels the logistical machine. Yet, the rituals of abstraction and valorization surrounding the production of its most prodigious export to China seek to reduce the logistical passage of commodities between the countries to an economic transaction. Indeed, a case can be made for understanding how logistical operations shape politics and economy in contemporary Chile through the frame of postcolonial capitalism—with an emphasis on questions of debt, crisis, extraction, and austerity—in much the same way as Ranabir Samaddar does for Greece (Samaddar 2016). This means understanding postcolonial capitalism as a global predicament that continues to be refracted through variegated spatial, institutional, and political conditions, rather than as a specifically subcontinental phenomenon.

In the end, the question that haunts this chapter and volume is not about whether understandings of capitalism inflected through either the Chinese or Indian experience are applicable to distant locations. Investigating the dynamics of region-making through the frames of labour and logistics offers a way of subtracting traditional notions of civilization and geopolitics from the debate about capitalism's variegations and regional forms. The emphasis falls rather on the materiality of infrastructural connections, the evolution of large technical systems, hardware arrangements, software codes, and patterns of interoperability. In the case of the sites investigated in the Logistical Worlds project, this focus extends the parameters of Asian regionalism beyond their usual boundaries. The Asia in Europe or the Asia in Latin America become just as relevant as the Asia in China or India. Yet, because logistics is not merely a means of organizing commercial activities but also a political technology that produces subjectivity and power, this patterning of space assumes a properly political form. Aside from the interests of states or regional organizations, therefore, such expressions of polity and their associated media forms define the limits and connective properties of capital's operations. While it is important not to reduce capitalism to logistics—and to measure its transformations, mutations, and effects against those produced by other prominent fields of capitalist activity, such as finance and extraction—these are the circuits of infrastructure, software, and labour that make logistical worlds.

REFERENCES

Allen, Robert C. 2003. *From Farm to Factory: A Reinterpretation of the Soviet Industrial Revolution.* Princeton: Princeton University Press.

Bear, Laura. 2015. *Navigating Austerity: Currents of Debt Along a South Asian River.* Stanford: Stanford University Press.

Bloch, Ernst. 1977. 'Nonsynchronism and the Obligation to its Dialectics.' *New German Critique* 11 (Spring): 22–38.

China Daily. 2017. 'Piraeus Port Has Never Witnessed Such Glory: PCT Employee.' 15 May 2017. http://www.chinadaily.com.cn/business/2017-05/15/content_29347561.htm

Collier, Stephen J. 2011. *Post-Soviet Social: Neoliberalism, Social Modernity, Biopolitics.* Princeton: Princeton University Press.

Cowen, Deborah. 2014. *The Deadly Life of Logistics: Mapping Violence in Global Trade.* Minneapolis: University of Minnesota Press.

Easterling, Keller. 2014. *Extrastatecraft: The Power of Infrastructure Space.* New York: Verso.

Gregson, Nicky, Mike Crang, and Constantinos N. Antonpoulos. 2017. 'Holding Together Logistical Worlds: Friction, Seams and Circulation in the "Global Warehouse."' *Environment and Planning D: Society and Space* 35, no. 3: 381–98.

Himatsingka, Anuradha. 2017. 'Kolkata Port Trust Records Highest Container Throughput in FY 2016-17.' *The Economic Times,* 14 March 2017. http://economictimes.indiatimes.com/industry/transportation/shipping-/-transport/Kolkata-port-trust-records-highest-container-throughput-in-fy-2016-17/articleshow/57632702.cms

Hatzopoulos, Pavlos, Nelli Kambouri, and Ursula Huws. 2014. 'The Containment of Labour in Accelerated Global Supply chains: The Case of Piraeus Port.' *Work Organisation, Labour and Globalisation* 8, no. 1: 12–28.

Home, Andy. 2017. 'China Still Hungry for Copper, But Not in Refined Form.' *Reuters.* 28 April 2017. http://www.reuters.com/article/China-copper-ahome-idUSL8N1HZ6S1

Kambouri, Nelli. 2014. 'Dockworker Masculinities.' *Logistical Worlds: Infrastructure, Software, Labour* (blog), 15 December 2014. http://logisticalworlds.org/blogs/dockworker-masculinities#more-279

Kanngieser, Anja, Brett Neilson, and Ned Rossiter. 2014. 'What is a Research Platform? Mapping Methods, Mobilities, and Subjectivities.' *Media, Culture and Society* 36, no. 3: 302–18.

Kim, Kap Hwan and Hoon Lee. 2015. 'Container Terminal Operation: Current Trends and Future Challenges.' In *Handbook of Ocean Container Transport Logistics: Making Global Supply chains Effective,* edited by Chung-Yee Lee and Qiang Meng, 43–73. Cham: Springer.

Kitchin, Rob and Martin Dodge. 2011. *Code/Space: Software and Everyday Life.* Cambridge, Mass.: MIT Press.
Labrut, Michele. 2015. 'West Coast Latin American Ports Expand Amid Fierce Competition.' *Journal of Commerce*, 4 December 2015. http://www.joc.com/port-news/south-american-ports/west-coast-latin-america-ports-expand-amid-fierce-competition_20151204.html
Mezzadra, Sandro and Brett Neilson. 2013. *Border as Method, or, the Multiplication of Labor.* Durham: Duke University Press.
Mitra, Iman Kumar and Mithilesh Kumar. 2016. 'Kolkata as a Logistical Hub with Special Reference to Kolkata Port.' *Policies and Practices – Logistical Spaces III: Hubs, Connectivity and Transit* 78: 17–33.
NAVIS. 2016. 'Terminal Pacífico Sur Valparaíso Goes Live with N4 Terminal Operating System.' 4 May 2016. http://navis.com/news/press/terminal-pacifico-sur-valparaiso-goes-live-n4-terminal-operating-system
Neilson, Brett. 2017. 'Belt and Road: State Transformation and Large Technical Systems.' *The Interpreter*, 15 May 2017. https://www.lowyinstitute.org/the-interpreter/belt-and-road-state-transformation-and-large-technical-systems
Olivier, Daniel and Brian Slack. 2006. 'Rethinking the Port.' *Environment and Planning A* 38, no. 8: 1409–27.
Parsanoglou, Dimitris. 2014. 'Trojan Horses, Black Holes and the Impossibility of Labour Struggles.' *Logistical Worlds: Infrastructure, Software, Labour* (blog), 15 December 2014. http://logisticalworlds.org/blogs/trojan-horses-black-holes#more-277
Quiroga, Javiera. 2017. 'Chile Hitches Ride on China Coattails as Copper Addiction Eases.' *Bloomberg*, 26 July 2017. https://www.bloomberg.com/news/articles/2017-07-26/chile-hitches-ride-on-China-coattails-as-copper-addiction-eases
Rojas, René. 2016. 'Out of the Ashes: The Resurrection of the Chilean Labor Movement.' *New Labor Forum* 25, no. 2: 36–46.
Rossiter, Ned. 2016. *Software, Infrastructure, Labor: A Media Theory of Logistical Nightmares.* New York: Routledge.
Samaddar, Ranabir. 2016. *A Post-Colonial Enquiry into Europe's Debt and Migration Crisis.* Singapore: Springer.
Selznick, Philip. 1949. *TVA and the Grass Roots: A Study of Politics and Organization.* Berkeley: University of California Press.
South China Morning Post. 2015. 'Port of Valparaiso Brings Chile and China Closer Together.' 30 September 2015. http://www.scmp.com/presented/business/topics/2015-chile-business-report/article/1862715/port-valparaiso-brings-chile
Spiliopoulou, Maria and Valentini Anagnostopoulou. 2017. 'Piraeus Port, Shining Star in Modern Maritime Silk Road.' *Xinhua*, 15 May 2017. http://news.xinhuanet.com/english/2017-05/15/c_136283017.htm

TCVAL. 2016. 'Trabajadores portuarios de Valparaíso y OPVAL firman importante acuerdo que establece mejoras laborales.' 8 February 2016. http://www.tcval.cl/sala-de-prensa/noticias/2016/08022016_trabajadores-firman-acuerdo-mejoras-laborales/

Tironi, Manuel and Israel Rodríguez-Giralt. 2017. 'Healing, Knowing, Enduring: Care and Politics in Damaged Worlds.' *The Sociological Review Monographs* 65, no. 2: 89–109.

Tsing, Anna. 2009. 'Supply chains and the Human Condition.' *Rethinking Marxism* 21, no. 20: 148–76.

World Cargo News. 2010. 'TSB Bags Piraeus Deal.' May 2010.

World Cargo News. 2013. 'TSB in the Process Mine.' May 2013.

INDEX[1]

A
Abstraction, 63, 151, 180, 283, 284, 286
Accumulation, 5, 7, 15, 17, 48, 50, 63, 64, 135, 222–237, 284, 285
Act East policy, 60, 285
Affective labour, 250, 254, 255
Afghanistan, 119, 128
Africa, 181, 182, 187, 236
Agriculture, 43, 93, 98, 103, 109, 147
Algorithms, 164, 167, 168, 177
Alternative media, 248, 249, 251–254
Amritsar-Kolkata growth corridor, 115
Andina, 282, 283
Andina mine, 282
Antonio, Negri, 244, 245, 250, 252, 255
Area studies, 8, 62, 63
Arms, 70, 81, 135
Army, 28, 41, 136, 137, 233, 235, 236, 247

ASEAN, *see* Association of South-East Asian Nations
Asia-led globalization, 4, 12–16
Asian Land Transport Infrastructure Development, 139
Asian Development Bank (ADB), 60, 61, 64, 188, 189
Asian Development Bank Institute (ADBI), 61, 62
Asian Highways (AH), 13, 138, 139, 186
Asian Infrastructure Investment Bank (AIIB), 4, 130, 179, 187–190
Asian Land Transport Infrastructure Development (ALTID), 187
Asian regionalism, 6, 8, 12, 286
Asian regions, 2, 12, 17, 223
Assam, 29, 47, 136–138, 141, 144
Association of South-East Asian Nations (ASEAN), 60, 61, 127, 128, 187

[1] Note: Page numbers followed by 'n' refer to notes.

© The Author(s) 2018
B. Neilson et al. (eds.), *Logistical Asia*,
https://doi.org/10.1007/978-981-10-8333-4

292 INDEX

Athens, 2, 263, 267
Austerity, 156, 159, 161, 163, 193, 274–276, 278, 284, 286
Australia, 27, 44, 71, 79
Authority, 41, 44, 47, 48, 54, 57–60, 63, 156, 193, 194, 276, 279, 285
Automation, 109, 184, 277, 282, 283
Azerbaijan, 175, 176

B
Bagdogra, 136–138, 142, 150
Baku, 175, 176, 194
Bangladesh, 8, 11–13, 52, 56, 62, 77, 114–120, 127–131, 136–139, 141, 142, 150, 189, 235
Barad, Karen, 203, 209
Basel Action Network (BAN), 199
Basel Convention, 207, 209
Bay of Bengal, 47, 53, 61, 62, 115, 117, 119, 120, 128, 130
Bay of Bengal Initiative for Multisectoral Technical and Economic Cooperation (BIMSTEC), 60
Bazaars, 142, 150, 200
Beijing, 129, 189
Bengal Chamber of Commerce, 29–31, 36
Bengal Initiative for Multisectoral Technical and Economic Cooperation (BIMSTEC), 128
Bharatiya Janata Party (BJP), 95
Bhutan, 12, 47, 114, 115, 128, 130, 136–139, 141, 142, 150
Bihar, 47, 73, 77, 80, 130, 140, 144
Biodiversity, 188
Borders, 49, 61, 74, 146, 162, 182, 204, 211–213, 231, 238
Border Security Force (BSF), 136
Bottlenecks, 61, 177, 179, 182–186
Boundaries, 2, 4, 16, 49, 182, 186, 286

Bratton, Benjamin, 3
British Indian Association, 33
Budge Budge, 25, 275
Burma, 27, 43, 44

C
Calcutta High Court, 71, 96
Calcutta Port Trust (CPT), 9, 23, 24, 55, 80, 81, 92, 93, 100, 110n1
Capital, 1–3, 6–8, 10, 13, 15–17, 49, 50, 54, 56, 62–64, 91, 103, 108, 110n1, 125, 136, 138–140, 142, 148, 151, 168, 175, 183, 202, 217, 222–232, 237, 238, 244–247, 249, 250, 258, 266, 273, 277, 278, 284, 286
Capital accumulation, 15, 222, 224, 225, 238, 246, 278
Capitalism, 2, 4, 6–8, 17, 49, 147, 168, 169, 180, 202, 216, 222, 225, 226, 229, 230, 238, 244, 246, 273, 285, 286
Capitalist accumulation, 3, 222, 224–226, 230–232, 237
Cargo handling, 9, 24, 39, 51, 53, 58, 95, 105, 121, 123, 265
Caspian Sea, 175
CATOS, 164, 170, 269–271
Central Asia, 119, 128, 176, 187, 188, 221, 223, 227
Centre of Indian Trade Unions, 100
Chandmoni Tea Estate, 146, 147
Chennai, 56, 121
Chicken's Neck, 12, 62, 136
Chile, 2, 263, 278, 279, 281–284, 286
Chilean ports, 280
China-Europe over, 176
China-India road, 114
China-Pakistan Economic Corridor (CPEC), 15, 128, 223, 224, 227–237

Chinese economy, 189, 224, 229
Chinese firms, 157, 158, 230
Chinese investments, 156, 157, 163
Chinese ports, 268
Chinese state, 2, 6, 221, 223, 229–231, 235, 237, 263, 267–268
Chinese state power, 228
Chinification, 14, 156, 160
Chitpore, 32, 39, 40
Chittagong, 56, 61, 62, 117, 118, 129
Chua, Charmaine, 3
Circuits, 4, 199–201, 206, 207, 211–213, 215, 286
Circulation, 1, 8, 12, 16, 17, 28, 49, 150, 179, 200, 203, 206, 210, 211, 215, 217, 225, 244, 245
Coal, 11, 27, 39, 43, 51, 92, 102–107, 122, 180, 274
Coal pellets, 102, 104, 105, 107
CODELCO, 282, 283
Cognitive capitalism, 168, 244
Colonialism, 5, 6, 156, 285
Commons, 104, 246, 252, 253
Commonsense, 179
Communist Party of India (Marxist) (CPI(M)), 81, 85, 95
Communities, 81, 84, 180, 226, 249–254, 257, 283
Computers, 54, 200, 205, 206, 283
Concession, 5, 6, 13, 155–158, 160, 162, 163, 171, 232
Conflicts, 8, 12, 16, 17, 52, 182, 217, 237
Connectivity, 4, 35, 60–63, 94, 115, 116, 122–131, 179, 181, 183–187, 190, 193, 199, 213, 221, 222, 224
Construction workers, 13, 149, 150
Consumption, 107, 148, 151, 165, 187, 225, 228, 230, 244, 249
Containerization, 9, 97, 192, 276

Containers, 8, 13, 69, 70, 79, 99, 100, 102, 108, 121, 124, 159, 162, 164, 165, 170, 209, 267, 271–276, 279, 283
  berths, 275
  operations, 274–276, 279
Contingencies, 9, 17, 24, 166
Contingency, 180
Contractual labour, 100
Control of labour, 162, 164–169, 171, 244
  mobility, 3
Copper, 208, 281–284, 286
Corporate social responsibility (CSR), 204
Corridor projects, 178, 181, 184, 188
Corridors, 6, 11, 12, 14, 15, 17, 62, 115, 130, 131, 176–195, 235, 244, 263–265, 272, 285
COSCO, 6, 13, 155–163, 171, 188, 263, 267–272, 285
Cossipore, 41, 58
Cowen, Deborah, 178, 189, 190
Cranes, 13, 27, 51, 99, 100, 120, 161, 164, 165, 170, 268, 270, 275, 280, 281
Crime, 9, 10, 17, 69–74, 76–88, 151, 209
Criminal activities, 78
Customs, 8, 34, 37, 44, 61, 170, 176, 178, 209, 228, 267, 279

D
Darjeeling, 139, 141
Darjeeling Himalayan Railway (DHR), 136
Data, 2, 3, 8, 11, 48, 54, 77, 78, 100, 162, 169, 170, 178, 184, 185, 210, 212, 213, 247, 267, 271, 281, 282, 284
  exchanges, 169, 170

Data (cont.)
  flows, 169–171
  mining, 167, 270
Day contracts, 280
Day labourers, 149, 280
Deleuze, Gilles, 157, 165, 166, 202
Desubjectivation, 158, 169
Digital networks, 169, 170
Disciplinary power, 164, 209, 210, 216, 217
Distribution centres, 162, 265, 278
Diversity, 180, 248, 253, 285
Dock
  labourers, 51, 76, 78, 82, 83
  workers, 53, 76, 83, 268, 276, 280, 283
Dock Labour Board (DLB), 82, 83, 95
Domestic work, 107, 109
Dredging, 52, 53, 92, 94, 114, 116–119, 125, 276
Duncan Group, 145, 148

E
East Asia, 3, 8, 114, 127, 229, 230
East Pakistan, 77, 114, 136, 235
East Pakistan/Bangladesh, 140
Easterling, Keller, 3, 175, 270
Eastern Bengal Railway, 31, 37, 39, 40
Economic crisis, 15, 62, 161, 268
Economic zones, 6, 17, 57
Eden channel, 95, 125, 126
Electronic data interchange (EDI), 170, 267, 271
Electronics, 201, 205, 210, 217, 282
Electronic waste, 199
Empire, 6, 17, 24, 27, 28, 32, 113, 189, 244, 285
Employment, 78, 100, 102, 103, 107, 110, 142, 147, 157, 160, 204, 236, 266
Enclaves, 6, 15, 192, 194, 235

Europe, 4, 115, 139, 156–158, 162, 163, 176, 177, 182, 183, 187, 191, 221, 227, 247, 264, 267, 269, 286
European Commission, 191, 268
European Union, 14, 176, 179, 207, 272
Event logs, 270, 271
Everyday logistics, 9, 24
E-waste, 14, 199–202, 204–217
  recycling, 12, 14, 15, 200, 204, 217
Exploitation, 149, 217, 245, 246, 250, 255, 275, 277
Extraction, 5, 7, 10, 11, 57, 192, 194, 206, 277, 282, 284, 286
Extrastatecraft, 175, 270

F
Facebook, 248–251, 253, 254, 256, 257
Failure of governance, 201, 208, 214
Feminization, 109, 160, 161
Feminized labour, 160
Fieldwork, 3, 13, 96–102, 110, 156, 248–256
Figurations of labour, 165, 166
Financial flows, 14, 179, 266
Financialization of space, 9, 48, 50, 56, 57, 59
Five Star Logistics, 94, 109
Fixed capital, 224, 230
Flesh trade, 143, 145
Foucault, Michel, 209, 211
Free labour, 246, 255–257
Frictions, 8, 177, 189, 194, 233, 266, 271

G
Gabrys, Jennifer, 204, 206, 213
Gadkari, Nitin, 57, 118, 119
Garden Reach, 27, 36, 48, 58, 65n1, 72–75, 77, 81, 120

INDEX   295

Gender, 4, 5, 9, 10, 14, 17, 84–86, 91, 97, 105, 160, 161, 172, 268
Geopolitical analysis, 15, 222, 223, 227
Geopolitics, 11, 60, 63, 114, 127–131, 222, 223, 231, 237, 238, 264, 286
Global China, 15, 223
Global economy, 183
Global flows, 97
Globalization, 1, 3, 4, 15, 17, 60, 114, 157, 158, 221, 224, 237, 238
Government of India, 35, 52, 56, 92, 108, 122, 124
Great Calcutta Killings, 80
Greece, 2, 14, 156, 159, 161, 188, 267, 268, 286
Guattari, Félix, 157, 165, 202
Gwadar, 11, 114, 128, 233
Gwadar port, 223, 232

H
Haldia, 11, 48, 51–53, 56, 61, 62, 65n1, 91–104, 107–110, 110n2, 116–120, 122, 123, 125–127, 129, 130, 264, 273, 274
Haldia Bulk Terminals (HBT), 95
Haldia Development Authority (HDA), 92, 97, 101, 102
Haldia Dock Complex (HDC), 52, 65n1, 92, 94–102, 105, 107–109, 122, 123
Haldia docks, 114, 117, 121
Haldia Integrated Development Agency Limited (HIDAL), 93
Haldia Law College, 110n2
Haldia Municipality, 93, 98, 100
Haldia Petrochemicals Limited (HPL), 93
Haraway, Donna, 156, 157, 161
Hardt, Michael, 244, 245, 250, 252, 255

Harvey, David, 48, 225, 231
Hegemony, 49, 216, 235, 249
Hinterlands, 12, 39, 47, 55, 92, 102, 107, 114, 121, 130, 135, 263–265, 272, 273
development, 222
Hong Kong, 4, 12, 14, 15, 142, 199–217, 223, 243, 264
Hooghly, 11, 23, 24, 32, 35, 39, 47, 48, 51, 53, 54, 91–93, 95, 114, 122, 124, 125, 264, 272, 274
Howrah, 26, 34–36, 38, 39, 44, 84, 85, 137, 143

I
Immaterial labour, 15, 16, 244–247, 249, 251–258
Immigrants, 70, 77, 82, 86, 140
workers, 83
Imperialism, 6, 28, 64, 171
Indian government, 12, 51, 56, 117–119
Indian Oil Corporation (IOC), 93, 94, 99
Indian ports, 56, 116
Indian Ports Association, 57, 65n2
Indian Tea Association, 37, 148
Inequalities, 16, 226, 267
Information, 1, 15, 16, 73, 91, 103, 110, 143, 169, 170, 180, 199, 201, 208, 209, 213, 214, 244–250, 252, 257, 267, 272, 273, 280
Information Handling Services (HIS), 176
Information society, 247
Informational capitalism, 246
Infrastructure projects, 93, 117, 221, 225, 227, 232
Insecurity, 69, 70, 76, 86, 88
International Monetary Fund, 188, 268

Internet, 12, 15, 16, 168–170, 199, 244, 246–247, 249, 252–256, 258, 272
immaterial labour, 246
users, 16, 244, 246–251, 255–257
Interoperability, 8, 170, 179, 186, 267, 271, 273, 285, 286

J
Japan, 4, 5, 28, 117, 127–129
Jawaharlal Nehru Port Trust, 119, 121
Jute, 29–31, 34, 35, 39, 81

K
Kahlon, R. P. S., 70, 71, 114, 117
Kandla port, 119, 121
Khalili, Laleh, 3
Kidderpore, 9, 26, 32, 34, 51, 54, 65n1, 80, 82, 84, 264, 272, 273, 275–277
Kolkata docks, 120, 121
Kolkata Dock System (KDS), 47, 65n1, 70, 122, 275
Kolkata port system (KoPS), 11, 114, 116–121, 126, 274, 276
Kolkata Port Trust (KoPT), 10, 11, 47, 48, 51–60, 62–64, 65n1, 70, 71, 84, 93, 95, 96, 105, 108, 110n1, 114, 116, 117, 119–121, 123–126, 272–275, 278
Kunming, 11, 115, 116, 130, 131

L
Labour
histories, 171
intensification, 277
mobility, 5
processes, 157, 164, 168, 269, 280, 284
regimes, 3, 8, 158, 169, 264, 265, 271, 272, 276, 284, 285
relations, 13, 156–158, 171, 268, 271, 276
rights, 159–162
subjectivities, 158, 245
unions, 53, 158, 160
Land use, 57–59
Late capitalism, 244
Latin America, 279, 286
Lazzarato, Maurizio, 165, 168, 169, 244, 252
LeCavalier, Jesse, 2
Lefebvre, Henri, 25, 193
Links, 176
Logistical coordination of production, 265, 266
Logistical expansion, 4, 5, 263, 281, 285
Logistical governance, 9, 13, 14, 48, 50, 52, 59, 60, 163, 178–182, 270
Logistical infrastructure, 15, 61, 130, 179, 187, 244, 246
Logistical networks, 11, 78, 91, 94, 96
Logistical operations, 7, 9, 14, 53, 97, 107, 157, 161, 171, 177, 178, 186, 192, 194, 195, 223, 227, 238, 265, 266, 273, 277, 278, 285, 286
Logistical politics, 50, 192, 193
Logistical power, 14, 177, 189, 193, 195, 226
Logistical practices, 2, 12, 15–17, 200, 201, 206, 213, 217, 265, 269, 272, 274, 278, 284, 285
Logistical revolution, 190, 191, 193
Logistical space, 10, 11, 79, 97, 99, 102, 226
Logistical systems, 91, 266, 273, 277, 278, 284
Logistical technologies, 157, 169, 273
Logistical Worlds project, 110, 264, 284, 286
Logistical Worlds research, 265, 268, 278
Logistics industries, 2, 266

M

Macau, 15, 16, 243, 244, 246–251, 253–255, 257, 258, 264
Machinic assemblage, 13, 157, 160, 163–168, 171, 172
Machinic idleness, 167–171, 270
Machinic subjectivities, 169
Manual labour, 41, 236
Maritime Silk Road, 187
Marwaris, 141, 142
Marx, Karl, 226
Marxism, 5, 7, 244
Massumi, Brian, 212, 214, 215
Materiality, 14, 15, 178, 200, 202, 203, 208, 223, 286
Migrants, 76, 102, 140, 141, 159, 161, 215, 231
  workers, 9, 276
Migration, 102, 104, 135, 141, 211, 228
Militarization, 3, 15, 226, 227, 235–237
Military, 13, 41, 62, 79, 136, 137, 150, 228, 231, 233–237, 280
  securitization, 150
Ministry of Surface Transport, 65n1
Mitchell, Timothy, 44, 48, 49
Mobilities, 164, 170, 200, 204, 209, 211–213
Modi government, 119, 131
Money, 16, 61, 71, 72, 87, 91, 98, 128, 142, 144, 145, 149, 184, 266, 277
Mumbai, 57, 101, 143, 144
Myanmar, 11, 56, 61, 62, 114, 115, 119, 127–131

N

National Highway Authority of India (NHAI), 118
NAVIS, 8, 164, 170, 269, 280
Negri, Antonio, 255, 256
Neoliberal capitalism, 49
Neoliberal development, 139, 145, 147, 149–150
Nepal, 12, 47, 82, 114, 115, 128, 130, 137–139, 142, 150
Netaji Subhas docks (NSD), 272, 275
New Silk Road, 176, 181, 187, 223
NGOs, 145, 201, 207, 217
La nombrada, 280, 281, 283, 284
Northeast India, 115, 135, 137, 139

O

Ong, Aihwa, 3
Online communities, 250, 252
Operating systems, 164, 269
Optimization, 57, 158, 162, 164–168, 170, 280
Original Design Manufacturers (ODM), 205
Original Equipment Manufacturers (OEM), 205
Over-accumulation, 15, 224–231, 237

P

Paira port, 119, 129
Pakistan, 8, 11, 15, 114, 115, 119, 127, 128, 187, 189, 223, 224, 227, 232–238
Panama Canal, 116, 278, 279
Parameters, 125, 185, 211, 274, 286
Peer production, 252, 253
Pipelines, 11, 17, 94, 119, 130
Piraeus Container Terminals (PCT), 13, 156, 158, 159, 161–167, 169, 171–172, 268, 269, 271
Piraeus port, 155, 156, 158, 164, 166, 167, 169, 263, 270–272
Piraeus Port Authority (OLP), 155, 156, 158–160, 164, 268, 269, 271, 272
Police, 30, 70–76, 80, 82, 83, 85, 87, 136, 143, 144, 147
Political economy, 87, 151, 183

Political geography, 234, 235
Political order, 11, 12, 17
Political spaces, 16, 177, 270
Port city, 11, 81, 91, 97, 113
Port Commissioners, 25, 27, 29, 30, 33–37, 39–41, 44
Port logistics, 71, 78, 105
Postcolonial capitalism, 5–12, 285, 286
Poverty, 64, 77, 107, 110, 140, 161, 229
Precarious labour, 13, 156, 163, 171
Precarity, 13, 159, 160, 168, 171, 271, 275
Precarization, 158, 172, 268
Primitive accumulation, 5, 147, 226
Privatization, 120, 155, 156, 163, 268, 275, 276
Process mining, 8, 270, 271
Production
 networks, 1, 179, 186, 265, 282
 processes, 164, 177
 of space, 16, 232, 273
 of subjectivity, 16, 17, 245
Productivity, 3, 16, 161, 162, 165, 170, 182–186, 192, 212, 269, 274
Protocols, 160, 161, 168–170, 177, 187, 207, 271, 281
Public-private partnership (PPP), 56, 183, 275
Punjab, 47, 233–237

R
Railways, 24, 28, 32, 35, 37, 39–41, 43, 44, 50, 92, 109, 126, 180, 265, 279
Real estate, 108, 146, 147, 180, 236, 274
Regional cooperation, 61, 178
Regionalism, 2, 62, 264
Regions, 2, 3, 8, 12, 63, 86, 127, 130, 135, 137, 140, 179, 180, 192, 200, 213, 222, 224, 228, 231, 233, 235–237, 247
Rent, 5, 10, 33, 57, 59
Repair, 40, 49, 58, 109, 161, 171, 200, 201, 203, 205
Research platforms, 264
Resistance, 17, 64, 166, 177, 194, 226, 231, 278
Reverse logistics, 201, 204, 213, 214, 217
 industries, 213
Rise of China, 5
Riverine port, 11, 52, 76, 93, 94, 114, 117, 121, 122, 131
Roads, 11, 25, 39, 43, 61, 92–94, 107–109, 135, 138, 143, 161, 176, 190

S
Sagar Island, 12, 47, 53, 62, 95, 116–118, 124, 126, 127
Sagar port, 118, 124, 126
 project, 118
Sagarmala project, 57
San Antonio, 279, 281
Sandheads, 25, 53, 56, 95, 120, 123
Sardari system, 276, 277
Sassen, Saskia, 192
Securitization, 3, 88, 137, 150, 226, 236, 238
Security, 9, 10, 13, 24, 31, 37, 69–71, 75, 78, 79, 85–88, 115, 117, 150, 158, 159, 211, 212, 217, 228, 236, 267
Seth, Laxman, 98, 100, 102
Seth, Tamalika Panda, 98, 99
Sikkim, 136–139, 142
Siliguri, 12, 13, 62, 85, 135–146, 149–151
Siliguri Corridor, 12, 136, 137
Singapore, 4, 8, 79, 82, 83, 121, 129, 186, 272, 273, 275, 276

Sittwe port, 62, 129
Smooth space, 8, 15, 223, 237
Smuggled goods, 10, 78, 82, 151
Social cooperation, 7, 277, 284
Social media, 246, 250, 252–255
Social space, 15, 25, 195, 222–224, 226, 227, 231, 233–238
Soft infrastructure, 14, 178, 187
Software, 2, 3, 8–10, 13, 15, 16, 54, 61, 156, 164–167, 169, 170, 175, 180, 200, 201, 205, 213, 214, 217, 244, 263–274, 277–284, 286
Software systems, 16, 157, 170, 267, 269, 271–274, 279, 282, 284
Sonadia, 114, 115, 117, 119, 128–130
South Asia, 4, 8, 26, 62, 86, 127, 130, 135, 151, 223, 227, 232
South-East Asia, 5, 13, 15, 23, 28, 56, 60–62, 71, 83, 116, 119, 127, 129, 130, 135, 139, 140, 222, 243
Sovereign state, 159, 160
Spatial fix, 15, 222, 224–227, 230, 231, 235–238
Spatialization of calculability, 9, 48, 50
Sri Lanka, 8, 11, 114, 125, 128, 273
Standards, 9, 14, 61, 179, 182, 185, 186, 210, 228, 267
State power, 222, 223, 227, 228, 238
State space, 222, 233
Statistics, 85, 140, 141, 147, 148, 210, 269
Strand Road, 23, 24, 29, 31
Struggles, 16, 127, 158–161, 166, 171, 172, 267, 279, 280, 283, 284
Subcontracted labour, 166
Subjectivation, 158, 168
Subjectivities, 11, 103, 150, 151, 159, 160, 168, 171, 195

Supply chains, 1, 3, 14, 82, 109, 167, 177–182, 185, 188, 191, 192, 194, 205, 213, 214, 216, 265, 266, 271, 278, 282

T
Tajpur, 116, 118, 126–129
Tea, 13, 23, 29, 31, 36, 37, 51, 104, 107, 135–137, 142, 145–150
  gardens, 145–149
  workers, 146, 147, 149
Terminal operating systems (TOS), 8, 165, 267, 271, 284
Terranova, Tiziana, 168, 244, 255
Territorial control, 6, 201, 203
Territorial state, 192
Territory, 5, 6, 9, 12, 16, 24, 29, 130, 136, 156, 158–163, 178, 189, 194, 200, 204, 211–213, 216, 217, 231
Thailand, 61, 128
Tibet, 11, 114, 136, 138, 142, 231
Tides, 52, 53, 92
Timber, 13, 137, 275
Total Soft Bank Ltd (TSB), 8, 164, 166, 167, 170, 269, 270
Toxicity, 208, 209, 216, 284
Trade, 2–4, 7–9, 13–15, 23–29, 32–37, 39, 43, 44, 56, 60, 69, 70, 87, 113–115, 118, 119, 127, 129, 130, 136, 138, 141, 142, 145, 146, 149–151, 176, 178–180, 183–185, 188, 199, 200, 207, 209, 213, 221, 223, 243, 263–265, 270, 272, 279, 280, 282
Trade unions, 95, 97, 100, 158, 160, 171, 280, 283
Traditional media, 248, 249
Trafficking, 13, 135, 137, 138, 143–145, 149

300 INDEX

Trans-Asian Railway (TAR), 114, 139, 186
Transloading, 56, 95, 123
Transnational flows, 15, 69, 70, 237
Transport, 118, 129, 139, 192
  infrastructure, 61, 126, 128, 223
  systems, 186
Transport Corridor Europe-Caucasus-Asia (TRACECA), 176
Transshipment, 8, 175, 267, 272, 279
Trucks, 11, 13, 94, 100, 102, 105–107, 109, 165, 167, 275, 276
Tsing, Anna, 165, 180

U
Unemployment, 78, 142, 161, 243
United Nations Economic and Social Commission for Asia and the Pacific (ESCAP), 139
United Nations Economic and Social Commission for Asia and the Pacific (UNESCAP), 186

V
Valorization, 255, 283, 284, 286

Valparaíso, 2, 16, 263, 267, 272, 278, 279, 281, 282, 284, 285
Valparaíso port, 263, 279, 281
Ventanas, 282, 283
Vercellone, Carlo, 168
Vessel Traffic Management System (VTMs), 126
Vietnam, 127
Virno, Paolo, 244

W
Warehouses, 9, 11, 23–26, 28–38, 41, 44, 51, 57, 94, 97, 100, 109, 179, 200, 265
Waste regimes, 203, 204
West Bengal government, 96, 118, 126, 137
Women, 13, 75, 77, 82, 85–87, 91, 97, 100, 102–110, 144, 145, 157, 159, 161, 215, 235, 276
  labour, 104, 107
World Bank, 61, 178–180, 188, 189
World economy, 193, 228, 229
World regions, 3, 16, 258, 285

X
Xinjiang, 231–233